Communicating European Research 2005

Communicating European Research 2005

Proceedings of the Conference, Brussels, 14–15 November 2005

Edited by

Michel Claessens
European Commission, Brussels, Belgium

Springer

A C.I.P. Catalogue record for this book is available from the Library of Congress.

ISBN-10 1-4020-5357-6 (HB)
ISBN-13 978-1-4020-5357-3 (HB)
ISBN-10 1-4020-5358-4 (e-book)
ISBN-13 978-1-4020-5358-0 (e-book)

Published by Springer,
P.O. Box 17, 3300 AA Dordrecht, The Netherlands.

www.springer.com

Printed on acid-free paper

TABLE OF CONTENTS

SCIENCE EDUCATION

TELEVISION

RADIO

MEDIA AND PRESS

SECTORAL COMMUNICATION

CONCLUSION

ACKNOWLEDGEMENTS

The editor would like to thank first all the authors who have contributed to this book. I would like to thank my colleagues of the Information and Communication Unit of the Directorate-General for Research of the European Commission for their help and support. I am in particular grateful to Hazel Collier, who did a great job in editing all the articles contained in this book.

The authors were all speakers at the "Communicating European Research 2005" conference which was organized by the European Commission in Brussels on 14 and 15 November 2005. More information on the Conference, including the programme and speakers' presentation, is available on the web site: *http://europa.eu.int/comm/research/cer2005.html*

The conference was the first ever organised by the Commission on *communicating research*. It shows the growing importance and recognition of public communication of science and technology.

The conference was a major success, as illustrated by the sheer number of participants (2,100+), the number of sessions and the quality of the speakers. The present book is a logical extension of such an original and enriching event.

Brussels
September 2006

LIST OF AUTHORS

Arata, Manuela is the head of the Office for Promotion & Collaboration Development within the Directorate General of CNR. From 1995 till June 2005 she was General Director of the INFM. Since 2003 she chairs the Associazione Festival della Scienza. Since 2003 she is involved as an independent expert in the evaluation of projects of the EU. She is a member of the Scientific Council of the Business School of Milan Technical University. She was awarded the AIDDA prize 2004. In March 2005 she was decorated by the President of the Italian Republic of the Award of Merit of the Italian Republic.

d'Andrea, Luciano, sociologist, is senior researcher at the CERFE, Roma. His work is focused on issues related at once to social dynamics and economic and technological transformations (innovation processes, scientific and technological research, health and urban development, labour market). Presently, he is coordinator of a research and training project on university spin-off. Recently, he wrote a handbook on socialisation processes in scientific and technological research as well as various articles on social dynamics in science and innovation and on scientific communication.

Belcher, Teresa is communications director for Med-Vet-Net, an EU-funded network of Excellence on zoonoses research. Teresa completed a BSc (Hons) in biological and environmental sciences at Murdoch University in Western Australia, and worked for a number of years in bioremediation, waste management and environmental consultancy. Following the completion of a Masters in science communication at the Australian National University, Teresa has worked in event management, education, journalism, PR/communications and corporate communications in Australia, Switzerland and the UK. In her role in Med-Vet-Net, she manages and provides a focal point for internal and external communications which includes the website, newsletter, publications, media, public awareness, and training interns in science communication.

Bennett, David J. has a PhD in biochemical genetics and an MA in science policy studies with long term experience, activities and interests in the relations between science, industry, government, education, law, the public and the media. He has worked in universities and companies in the UK, USA, Australia and The Netherlands. He is a member of numerous national and international organisations and committees in biotechnology, and works with the European Commission, government departments, companies, universities, public interest organisations and the media in these areas.

Bettencourt-Dias, Mónica is a research associate at the University of Cambridge, where she studies cell biology of cancer. She did her PhD in Cell Biology at the University College London and a Diploma in Science Communication at Birkbeck

College (UK). She co-organises workshops with the aim to train researchers in communicating skills and co-edited a booklet on how to communicate science to lay audiences. She is a founding member of the non-governmental organization, Science for Development, which aims to promote science and technology in developing countries. She will start her own laboratory on the cell biology of cancer in October 2006 at the Gulbenkian Institute of Science (Portugal).

Børresen, Torger, PhD, is the Research Director of the Department of Seafood Research at the Danish Institute for Fisheries Research, and professor of Food Technology at the Technical University of Denmark. He has a long-standing experience as manager of research at all levels. Due to his academic skills he has been serving at research councils, advisory groups for research within the food and technology area nationally and internationally, and is still active in the academic world, serving e.g. on editorial boards for scientific journals. He has a long experience in collaborating with industry, and has an extended network within both academia and among decision makers nationally and internationally. Torger is the co-ordinator of the IP SEAFOODplus.

Bufon, Milan is Associate Professor of Human and Political Geography and Head of the Department of Geography at the Faculty of Humanities of the University of Primorska (Koper, Slovenia). He is also the Director of the Institute for Mediterranean Humanities and Social Studies at the Science and Research Centre of the University of Primorska and is leading several national and international research projects concerning convergence and divergence social and spatial processes in Europe, and particularly in the Upper Adriatic. He is currently Vice-Rector for research at the University of Primorska and President of the Slovene Research Institute in Trieste.

Carpenter, Joanne co-ordinates the RELAY: Communicating Research programme at Panos London, within a global network which stimulates debate on development issues. RELAY works with broadcast and print media to promote inclusive and informed debate on development research aimed at policy makers and wider publics in developing countries. Joanne has worked for a variety of UK non-governmental organisations, specialising in social development. She has experience of working with partners in South Asia and Africa to support rural community development. She spent time in Serbia conducting participatory action research to ensure the active inclusion of marginalised ethnic groups into relevant policy processes.

Ceva, Elena is a journalist, working for RAI in Milan (Italy). She graduated in Politics, Philosophy and Economics at Balliol College, University of Oxford (U.K.). She has been working for the last ten years for the sciences and medicine news desk, both at local and national levels.

Cheng, Donghong has been the Executive Secretary of China Association for Science and Technology (CAST) since June 2001. She has been Director-General of the Department of Science Popularization of CAST from 1999 to 2003 and

Director-General of the Department of Children & Youth Affairs of CAST from 1984 to 1999, where she has initiated and led many projects on science communications. She is now taking the leadership of many national initiatives, such as the Head of Office of the National Scheme for Scientific Literacy, and is the national focal point of the Asia-Pacific Network on Gender, Science and Technology of UNESCO. She is also a Member of the Executive Board of All China Women's Federation and a Board Member of the China Association for Science Instructors.

Chevassus-au-Louis, Nicolas turned to scientific journalism, after completing a PhD in neuroscience. As a science journalist he focuses mostly on policy science issues, and historical investigation. Among his last studies, are a pioneering study about scientific life in France during the Second World War (Savants sous l'Occupation, Le Seuil, Paris, 2004) and a reflection about machine breaking as a means toward the democratization of technological choices (Les briseurs de machines. De Ned Ludd à José Bové, Le Seuil, Paris, 2006)

Chreiki, Eric is chairman and co-founder of Innov.Europe, an independent observatory of innovation and information technologies. This organisation publishes reports on European IST innovators, studies on innovation trends, analysis, how-to guides and directories, specially focused on SMEs. Eric has worked over ten years both in the IT and the business press. He was chief editor (or associate CE) for leading IT magazines: Silicon.fr, 01 Informatique, Newbiz, PC Professionel, PC Magazine, Portable Magazine. Eric holds an MBA and degrees in Economy and Journalism. He now acts as an analyst and consultant for European innovative SMEs within the IST sector.

Chryssochoidis, George is a Business Administration graduate of the University of Piraeus and has a D.E.S.S. from the University of Paris II (France) in International Trade, an M. Phil from the University of Bath (UK) in Management and a Ph.D. from the University of Warwick (UK). He has lectured in British and French Universities. He is Expert in International Marketing from the International Centre of Trade (World Trade Organism/UNCTAD) of the United Nations and has participated in several national and EU funded research programmes either as contributor or as scientific coordinator. He teaches agri-food marketing, food marketing, new food product development, process development and innovation management.

Claessens, Michel has a Ph D in physical chemistry. He was researcher in physical chemistry, then in the biotechnology and the chemical industry, and has been a freelance scientific journalist. He joined the European Commission in 1994 and he is currently deputy head of the Unit information and communication in the Research Directorate-General. His main responsibilities concern the organisation of major conferences and the Eurobarometer surveys on science and technology. He is also the editor-in-chief of the RTD info magazine on European research and a writer.

Cnudde, Filip obtained his PhD in plant molecular biology in 2004 at the University of Nijmegen, the Netherlands. He is currently working as a post-doc at the Marketing and Consumer Behaviour Group of Wageningen University. In this position, he is the Dissemination and Communication Manager of the EU FP6 Integrated Project SAFE FOODS and a founding member of the FP6 Communication Managers Board COMMENT. In his work he integrates input from natural and social sciences in the communication of food safety issues.

Coërs, Pierre is in charge of internal and external communications for Solvay, a chemical and pharmaceutical international Group. He specialises in the areas of environment, health and safety. As a biologist and bioengineer, and with his experience in journalism, he focuses on societal questions regarding sustainable development, and on the various means of creating information flows and dialogue on industrial and science issues within society.

De Backer, Willy is editor-in-chief of EU policy portal EurActiv.com. He studied Germanic Philology at the Ghent University and Political Science at the Vrije Universiteit Brussels. He worked in the European Parliament as a political assistant from 1984 until 1993 and was co-Secretary-General of the International Coordination of Green Parties from 1985 until 1989. His main areas of political expertise are environment and sustainable development policies and energy.

De Saeger, Emile is leader of the Air Quality and Health action at the Institute for Environment and Sustainability of the DG Joint Research Centre in Ispra. He joined the Joint Research Centre in 1989 and helped form the European Reference Laboratory of Air Pollution. This laboratory has supported the development of European policy on air quality through the development, validation, standardization and harmonization of air pollution measurements in monitoring networks. His responsibilities also include air quality modelling and the assessment of health effects from air pollution.

de Selliers, Jacques is co-founder and general manager of GreenFacts, is an electrical engineer. He has worked for Schlumberger as a field engineer and later as a software R&D engineer. He then joined the SOLVAY Group where, among other projects, he created the Advanced Process Control department. In charge of providing information on environment and health issues and regulations, he felt the need for a better access to scientific information in this domain. He initiated and developed the GreenFacts concept and its three-level structure to ensure the accessibility of scientific information.

De Semir, Vladimir is Commissioner for Scientific Culture – Barcelona City Council, associated professor of Scientific Journalism and director of the research centre Science Communication Observatory at the Pompeu Fabra University (Barcelona), President of the international network Public Communication of Science and Technology (PCST), member of the European Network of Science Communication Teachers (ENSCOT) and the European Science Communication Network (ESCONET). He is also a journalist specialised on scientific and medical issues since 1982 and director of the magazine Quark (Ciencia, Medicina, Sociedad y Cultura).

Demeneix, Barbara is Director of the Department on regulation, development and diversity of molecules of the Museum national de l'histoire naturelle since Sept. 2002 as well as the Director of the UMR CNRS 5166 on the evolution of endocrine regulations since January 1998. In 2002 she received the national jury award for her "Watchfrog" project at the competition on "the creation of enterprises and innovating technologies" organized by the ministry for research and technology.

Demeter, András is a nature conservation policy officer in the DG Environment of the European Commission. With a degree in biology and a doctorate in zoology, he devoted the first years of his career to research on the taxonomy and ecology of mammals in the Hungarian Natural History Museum. He then became a science administrator in the head office of the Hungarian Academy of Sciences. Later, András Demeter was invited to head one of the nature conservation departments in the Ministry of Environment of Hungary where he formulated and implemented nature conservation policy.

Duane, Berta is press officer at the Joint Research Centre's (JRC) site in Ispra, Northern Italy. She has been working for 18 years at the JRC and gained a profound knowledge of the activities of the JRC in various positions. Since three years she works as press officer and exploits this background knowledge in her daily contacts with local, national and international media.

Duke, Sean has been a full-time science and technology journalist for the past ten years. He was educated at University College Dublin, where he obtained a B.Sc., and the New York University School of Journalism, where he achieved an M.A. in Science and Environmental Reporting. In 2003 he was appointed the editor of Ireland's leading technology magazine "Technology Ireland". Sean is also Joint Editor and founder of "Science Spin" Ireland's first and only science and discovery magazine. Recently, he co-founded another new magazine, this time for manufacturing and applied research in industry, called "Spin Industry".

Egikova, Viola is President of Association for Science Writers and Journalists Intellect, Russia. She was born in Georgia, where graduated State University, continued education in Moscow University. Since 1974 works as a science journalist in Russian media, is science desk chief in daily independent newspaper Moscowskaya pravda. President of the Association for Science Writers and Journalists Intellect and Russian representative in EUSJA. She works much with foreign colleagues for advancement of science journalism, organising study trips and meetings of science journalists. Was a panel member of Euroscience in Stockholm, conference of World Federation of Science Journalists in Budapest and Montreal.

Engelmann, Ulla is the Head of Unit of the Public Relations Division in the European Commissions' Joint Research Centre (DG JRC). She studied chemistry at university in Karlsruhe, Germany and completed a PhD at the National Research Centre in Karlsruhe. Ulla worked for a total of 11 years at the JRC Ispra site in research and strategic communication roles. Promoted to head of Unit, her work now centers on similar issues across the seven institutes of DG JRC.

Field, Robert works at present for the European Commission to develop understanding of population exposure to air pollution. After his bachelors degree in Resources

Science at Kingston University he completed a research PhD on urban air quality at Imperial College. Subsequently he worked on novel on-line data management procedures for air quality networks at AEA Technology. He then lectured at the School of Environment and Natural Resources of the University of Wyoming. His next position will be lecturing, in particular on the sustainability of environmental systems.

Gale, Carolyn wears many hats, which includes communication and e-learning consultant, designer, researcher, and trainer. After developing a university-level program to teach academic researchers how to communicate to lay audiences, she is currently co-founding an international research communication firm. She is also a Visiting Scholar at Stanford University (USA), continuing research into explanatory communication (how experts learn to communicate complex information) and teaching courses on writing, community-building, and blogging.

Gascoigne, Toss is Executive Director of the Council for Humanities, Arts and Social Sciences (CHASS); and prior to that held a similar position with the Federation of Australian Scientific and Technological Societies (FASTS). He has developed workshops for scientists in media skills, and communication planning, which they have run in several countries for 12 years. He developed and organised since 1999 big national meetings between scientists and federal parliamentarians in Australia, called "Science meets Parliament". This has brought hundreds of scientists directly in touch with federal parliamentarians.

Harding, Gareth is United Press International's Chief European Correspondent. After brief stints as a political advisor in the European Commission and European Parliament, he turned to reporting on EU affairs for the European Voice, Time Magazine and the Independent. A documentary film-maker and radio and TV pundit, he is also a course leader for the European Journalism Centre in Maastricht. He has written "Reporting the EU from outside Brussels" – a handbook for journalists. Gareth Harding has lived in Norway, the Czech Republic and Belgium since graduating from the London School of Economics in 1990.

Hayhurst, Richard is a leading life sciences PR consultant having founded the agency HCC De Facto. The agency's clients included Dolly the Sheep, Roslin Institute, Genzyme, CAMR and GSK and they dealt with issues such as stem cell research, cloning, environmental testing, developing world agriculture and science, gene therapy and genetic testing. Currently Richard runs his own agency Hayhurst Media with a variety of private and public sector clients across Europe including EuroGentest and Nanobio-RAISE.

Heckl, Wolfgang M. is the Director General of the Deutsches Museum and professor of experimental physics and nanotechnology at the University of München (LMU) with a special interest in the field of organic self-assembly and the origin of life. His academic teachers are Nobel Prize winners Gerd Binnig and Theodor Hänsch. As a dynamic and charismatic science communicator he received the Communicator Prize in 2002 from the German Science Foundation and was awarded the first European

Descartes Prize for Science Communication in 2004. He is the chairman for the next pan-European Euroscience Open Forum in July 2006.

Hodge, Russ is Public Information Officer at the European Molecular Biology Laboratory (EMBL) in Heidelberg. He is author of the EMBL Annual Report and has been instrumental in the development of science communication and education projects at EMBL, within the EIROforum, and at the Universities of Heidelberg and Ulm. He has written grants establishing the European Learning Laboratory for the Life Sciences (ELLS) and the international journal Science in School, and was chairman of the EIROforum's Working Group on Outreach and Education for two years. He has recently been awarded a contract to write a series of books about Genetics and Evolution for the public from a major New York publisher.

Hoegberg, Pi M.Sc. (nutrition) and Ph.D. (toxicology), initiated the network of FP6 communication managers. During 2004 and 2005, she was the Dissemination Manager of CASCADE, and EU-funded Network of Excellence on chemical contaminants in food, where she had chief responsibility for the joint dissemination and communication tasks. Between 1997 and 2003 Dr. Högberg conducted research at Karolinska Institutet, Institute of Environmental Medicine, on the effects of environmental toxins on retinoid (vitamin A) processes in vivo. At present, she is public health officer at the Swedish National Public Health Institute.

Holliman, Richard is Lecturer in Science Communication and the Public Understanding of Science at the Open University, visiting lecturer on the Birkbeck College Diploma in science communication and researcher in science and society at the Open University. He was a member of the European Union-funded European Network of Science Communication Teachers (ENSCOT). His primary research interest lies in examining representations of scientific knowledge in the public sphere. He is currently working on the Invisible Witnesses: scientific role models and stereotypes of women scientists on television project.

Hunt, Roderick is Visiting Professor in Biosciences at the University of Exeter. After studying at the University of Sheffield, his research has involved plant growth analysis, climate change, plant strategy theory, hierarchy theory, expert systems and cellular automata modelling. He currently helps manage both the Annals of Botany, the world's oldest general botanical title, and the European Association of Science Editors, an international group promoting the sharing of linguistic traditions and professional experience within the context of science communication and editing.

Jergović, Blanka holds a Ph.D. in political sciences/media studies. She is science journalist – editor on the Croatian Radio (Croatian Radio-Television) in Zagreb, lecturer in Media and Society Studies, University of Zagreb and University of Dubrovnik and Honorary Research Fellow at the University College London, Science and Technology Studies. She is principal investigator of the project 'The state of Science Communication in Croatia', researcher on European Science Communication Workshops (FP6) and Curriculum development of Science communication at University of Zagreb. She is co-director of the International science communication summer school in Dubrovnik. Her main scientific interest is science communication and media/society relationship.

Johansson, Jan-Olov works as senior editor on the Swedish Radio, where he started the award winning "Science Radio" that has three daily science news broadcasts on prime time. The science spot is one of the most popular among listeners and reaches around 1 million people nationwide in Sweden. He has also initiated many new feature shows about science in the radio and during his time as editor in chief Science Radio doubled it's airtime in the schedule. He is an environmental agronomist by training besides radio also writes columns in the press and has done TV for youngster.

Joy, Gill is Marketing Team Leader for IST Results a DG Information Society & Media service providing news on innovations emerging from EU-funded research in the Information Society Technologies area. An information and communications consultant for ESYS plc, Gill has seventeen years of experience developing, implementing and marketing new information systems at the research-industry interface. Gill's role in ESYS focuses on market analysis for emerging technologies and evaluation of the impact of ICT in research and education sectors. Her current work for IST Results involves promotion of the service to target audiences in enterprises, the research community, public authorities, investors, and the press/media.

Kikin-Gil, Ruth was born and raised in Israel, studied visual communications in Bezalel academy of art and design in Jerusalem, and later on lectured there. She co-founded an interactive design consultancy Max. Interactive. She moved in 2003 to Italy to pursue a Masters degree in Interaction design at Interaction Design Institute, Ivrea. She is working as an experience designer, consulting and developing new products. She is interested in the interplay between social behaviour and technology. She is also a guest editor at 3quarksdaily blog.

Lerch, Joachim is Chairman of Foerderverein Science und Technologie e.V. in Freiburg/Germany. This organisation is promoting science and technology to schools and the public. Since 2000 the Foerderverein organized a lot of Science Communication Events like Science Days and Science Festivals. Joachim Lerch was president of the European Science Events Association (EUSCEA) between 2001 and 2004. He studied Mathematics, Physics and Technology in Berlin and Freiburg.

Leshner, Alan is Chief Executive Officer of the American Association for the Advancement of Science and Executive Publisher of Science. Previously, Dr. Leshner had been Director of the National Institute on Drug Abuse at the National Institutes of Health, and Deputy and Acting Director of the National Institute of Mental Health. Before that, he held positions at the National Science Foundation, and was Professor of Psychology at Bucknell University. He is a member of the Institute of Medicine of the National Academies of Science, the American Academy of Arts and Sciences, and the National Science Board.

Lorenzen, Dirk H. was born in 1968 and has been a senior science reporter for German Public Radio and major newspapers since 1994. He graduated in astrophysics from Hamburg University. Lorenzen covers astronomy and space flight. He is the author of six books on these topics and has given hundreds of public talks. Lorenzen is a member of the Executive Board of the German Science Journalists' Association (WPK).

Maglio, Guglielmo has a degree in Natural Sciences (Università di Napoli Federico II) and professional experiences and activities in the field of science communication. He joined Città della Scienza in 1996, where, up to 2003 was responsible for Events and Science. He has collaborated with the Educational Department for the development of educational programs of Fondazione IDIS – Città della Scienza, and for the project of the permanent exhibition "Life Evolution". He was responsible for the European projects "Brain" and "Oceanics" and from 2004 is responsible for the thematic area Science and Society of the Fondazione IDIS. Today is involved in five European Projects on Science and Society (Nanodialogue, Cipast, DECIDE, ECD, EuEv).

Mazzonetto, Marzia has a degree in science communication and has attended the Master in science communication at SISSA, in Trieste, where she currently works as contract researcher with the Innovations in Science Communication research group. She has collaborated to the Scirab project, and currently works as Editorial Staff of JCOM, the Journal of Science Communication published by SISSA. In the past she worked for the national radio station Radio 24 II Sole 24 Ore, as free lance science journalist for the science news agencies Zadig and Galileo, in Rome, and for the CNR Press Office (Consiglio Nazionale delle Ricerche).

Merzagora, Matteo is a free-lance science journalist and teaches at the Master in science communication, SISSA, where he was manager of the project Science in Radio Broadcasting (SCIRAB). He co-produced and co-presented the weekly science radio programme "II ciclotrone" at Radio Popolare, of the daily "Le Oche di Lorenz" at RAI – Radio 3 and he is presently contributor to the daily programme "II volo delle Oche", Radio 24. He is contributor of several newspaper and magazines in Italy and France and has published four books on science communication in film, radio and museums.

Metcalfe, Jenni is Director of Econnect Communication, a small company specialising in science communication. Prior to this, she was a Communication Manager for CSIRO after working as a journalist. She is a founding member and current President of Australian Science Communicators, and a member of the Scientific Committee of the PCST Network. With Toss Gascoigne, she has run workshops in media and communication skills for scientists for the last 14 years.

Millar, Robin is Professor of Science Education at the University of York. He studied physics at university and completed a PhD, before training as a teacher. He then taught for 8 years in secondary schools. His work now centres on science teacher education, research and curriculum development. He has played a central role in the development of several innovative school science courses including Salters' Science, Science for Public Understanding and Twenty First Century Science. These courses explore how the school science curriculum can be given a stronger scientific literacy emphasis.

Miller, Steve trained in physical chemistry. During the 1980s he was a journalist for the Labour Party in London. He joined University College London in 1986, researching astronomy, before working on science communication. He researches public understanding of science, particularly the European dimension. With Jane Gregory, he wrote *Science in Public: communication, culture and credibility*. He directs the European Network of Science Communication Teachers *(http://www.enscot.eu.com)* and ESConet Workshops. He chaired the European Commission's *Expert Group Benchmarking the Promotion of RTD Culture and Public Understanding of Science (http://www.cordis.lu/era/benchmarking.htm)*.

Miles, Laura is responsible for the AlphaGalileo site strategy, development and operations both in the UK and across eleven partner countries in Europe. She represents science journalists, communicators and researchers in promoting effective dialogue between these groups and the public. She holds degrees in psychology (biological sciences) and science communication, and is a former science journalist, science public relations expert and multinational webmaster.

Mima, Noyuri is a Deputy Director of the National Museum of Emerging Science and Innovation, specializing in the field of Research, Development and Education. Prior to this post, she conducted researches on cognitive science and human computer interaction. Meantime, she served on several government committees, relating to science, technology and education. Additionally, she was involved in the establishment of the museum as a member of the planning committee. Currently, by utilizing her experience as a researcher and a learning theory in cognitive science, she is responsible for many aspects of the museum's implementation, working to build bridges between the public, researchers and the science policy makers.

Minoli, Luisa is responsible for EU Projects in API Varese – the Association of Small and Medium Enterprises of Varese Province – promoting the interests of, and providing services to, associated SMEs. She is involved in dissemination and stimulation actions, informing the associated enterprises about European R&D programmes, and promoting their participation in EU-financed projects. Luisa is the co-ordinator of the technical committee of "Innovare", the magazine of the Italian Confederation of SMEs, and author of numerous articles focused on European research.

Morris, Tara, with a doctorate in sciences, has eight years experience working with European Institutions in Brussels. He is currently editor-in-chief for the IST Results service, reporting on the fruits of EC funded ICT research projects, helping to bring emerging technologies closer towards business communities. Over these years he was actively involved in managing communication and awareness activities in the field of European R&D and policy, working primarily with DGs Information Society,

Research and Enterprise. He has hands-on experience in both print and electronic media at all stages ranging from conception, design, construction and maintenance.

Oehlenschläger, Jörg has a PhD in biochemistry. After some years at the University in Hamburg as assistant professor he joined in 1977 the Department for Seafood Research in Hamburg, Germany, where he is actually scientific director and is responsible for research on seafood quality, nutrition and seafood related legislation. As professor he is teaching technology of aquatic food products at two German Universities. He is chairing many international and national groups focussing on seafood quality and safety and is speaker of the German delegation to CCFFP since 1986. In the management team of the IP SEAFOODplus he acts as the dissemination and communication manager.

Pasveer, Bernike is assistant professor at the Faculty of Arts and Culture of the University of Maastricht. She studied sociology and completed her PhD in 1992. Her work focuses on issues of technology and the human body, old and new media, and of late also on (medical) technologies and research for development of the South. In 2005 and 2006 she worked at the Rathenau Institute on issues of political and public engagement with science and technology.

Pérez Ballesta, Pascual joined the DG Joint Research Centre of the European Commission in 1993, after completing his PhD in chemistry at the University of Murcia. Since then, his research at the European Reference Laboratory of Air Pollution was related to the monitoring of organic pollutants (BTEX, VOC & PAH). He has participated in numerous working groups for the development of European air quality legislation and contributed to the establishment of international standard methods. His current interest includes the development of monitoring approaches for population exposure to air pollutants.

Peters, Hans Peter is Senior Researcher at the Program Group Humans-Environment-Technology of the Research Center Jülich, Germany, and Adjunct Professor of Science Journalism at the Free University Berlin. His research deals with the formation of public opinion on science and technology under the conditions of a media society. In particular, he focuses on the interactions of journalists and scientific experts and on the impact of scientific knowledge on public meaning construction. He is member of the Scientific Committee of the International Network on Public Communication of Science and Technology (PCST).

Potočnik, Janez is the Commissioner responsible for Science and Research. He studied economics to doctoral level and has worked as an economic analyst in an agency before taking up a position as assistant director at the Institute of Macroeconomic Analysis and Development (IMAD) in his home country of Slovenia. From 1988 to 1993, he was senior researcher at the Institute for Economic Research and returned to the IMAD as director between 1993 and 2001. In 1998, he was head of the negotiating team for Slovenia's accession to the EU. Then, in 2000, he became acting director of the Government Office for European Affairs, joined the prime minister's cabinet in 2001 before taking up a post, in 2002, as minister for European Affairs. Janez Potočnik is a European Commissioner since May 2004.

Reding, Viviane has a Doctor of Human Sciences from Paris Sorbonne University. A native Luxemburger, she worked as a journalist for 20 years while at the same time pursuing her political career. She was a Member of the Luxemburg Parliament from 1979 to 1989 and subsequently served as a Member of the European Parliament until 1999. Mrs Reding was appointed European Commissioner for Education, Culture, Youth, Media and Sport in 1999 and is currently serving her second term, now as Commissioner for Information Society and Media.

Revuelta, Gemma is associated professor of Science Communication at the Universitat Pompeu Fabra (UPF). Assistant director of the Observatori de la Comunicació Científica (Science Communication Observatory) of the UPF. Head Editor of the journal Quark: Ciencia, Medicina, Comunicación y Cultura. Co-director of the Proyecto Quiral, from 1997 (medicine and health in the Spanish daily press). Lecturer and academical coordinator of the Master of Science Communication at the UPF. Member of the European Network of Science Communication Teachers (ENSCOT). She is the author or coauthor of several studies, specially focused on the media coverage of health and medicine, as well as on the relationship between scientific journals and mass media and science centres.

Ribele, Marta who was the news editor of "Dienas Bizness," Latvia's daily business newspaper, is now the head of the business news group of the major morning daily "Diena". Both newspapers are part of Bonnier group, one of the biggest Scandinavian media companies with subsidiaries in twenty European countries. Marta Ribele received the prize for outstanding achievements in journalism by Bonnier group and Latvian press publishers association in 1999. She holds a masters degree in social sciences from the University of Latvia. She is fluent in English and has a basic knowledge of French.

Roffi, Giuseppe, a physicist with a technological background, is director of the CNR-EU Liaison Office, the European office of the Italian Research Council in Brussels. Lately Head of international activities at CNR, he has strong connections with all European research institutions. He has pioneered networking of European scientific excellence, bilateral and multilateral cooperation in all fields of science including social and communication science. Scientific Attaché to the Italian Embassy in London for several years, he has developed considerable skills as promoter of scientific exchanges and has achieved interesting results in national and European cross-fertilization of academic and industrial research.

Roland, Marie-Claude is a linguist with a PhD in Didactics and Language Science she has specialized in the promotion of interdisciplinary research, developing a methodology based on reflective practice to improve scientists' communication skills, and helps develop effective and creative research teams. Her most recent interests extend to mediation and facilitation as means for creating new communication and collective learning modes among multiple stakeholders.

Schreiner, Camilla is educated as geophysicist and has recently completed a PhD in science education. She is now a postdoctoral researcher on the project ROSE (The Relevance of Science Education: see http://www.ils.uio.no/english/rose/) at the University of Oslo. Her research interest involves cross-cultural aspects of youth's attitudes and interests related to science and science education. She draws on sociological perspectives on youth culture for understanding Western youth's views on science and technology.

Sheridan, Cormac is a freelance science journalist specialising primarily in biotechnology. He is a frequent contributor to Nature Biotechnology and Nature Reviews Drug Discovery. He reports on the European biotechnology industry, particularly in Ireland, Scandinavia, Switzerland & Southern Europe, for the US biotechnology industry publication BioWorld International. He is also a contributor to Technology Ireland magazine and, occasionally, to the Irish Times. He is guest lecturer on genetics & media on the B.Sc Genetics course at University College Cork. Cormac is currently president of the Irish Science & Technology Journalists Association.

Sjøberg, Svein is Professor in science education at Oslo University, Norway. His research interests are the social, cultural and ethical aspects of science education, science education and development, gender and science education in developing countries, and issues of scientific literacy and public understanding of science. In 2005 he was awarded the price of the International Union of Pure and Applied Physics for outstanding contributions to physics education. He was member of the Advisory Group on the Science and Society action plan for FP6. Organizer of ROSE (The Relevance of Science Education), a comparative project in 40 countries on pupils' interests, attitudes etc. of importance to science teaching and learning.

Stipsits, Michaela is managing editor of the Austrian magazine "Business People" that is published every two months as a supplement to "WirtschaftsBlatt", the leading national business newspaper. "Business People" is a manager magazine with focus on Austria's leading companies, entrepreneurs and promising start-ups. As a freelance journalist for several years, she joined "Business People" in May 2002, in the position of managing editor. Her job is to order and discuss stories with the editorial staff and freelance writers, plan the issues, edit stories, and to look after deadlines. She is also still investigating and writing stories for the magazine.

Terry, Andrew is a Project Officer for the IUCN Regional Office for Europe in Brussels. He works with the Programme Coordinator to develop and implement IUCN's European Programme. Part of his role is to act as a focal point for IUCN activities relating to threatened species and Protected Areas. He is also the European Coordinator for IUCN's World Commission on Protected Areas. As a research biologist he worked at the University of Copenhagen on the evolution of animal Communication and the use of animal behaviour as a tool for monitoring populations in the wild.

Tola, Elisabetta is a lecturer in microbiology and has a Master in science communication. Besides writing for different magazines, she produces and presents science radio programmes and is currently a presenter for Radio 3 Scienza, the daily science programme of the Italian national radio, RAI. She also works as a communication consultant for different scientific institutions. She has been involved in SCIRAB, a European project aimed at building a network of science radio programmes throughout Europe. Together with other colleagues, she has recently started off a new science communication agency based in Bologna, Italy.

Trench, Brian is Head and senior lecturer of School of Communications, Dublin College University, member, scientific committee, Public Communication of Science and Technology network. Brian Trench has been an adviser and trainer on communication for community-based and non-governmental organizations. He has organized media and communication skills training workshops for academics, state agency employees, postgraduate researchers, and community organizations. Brian Trench's research interests are in science communication, online journalism and social uses of technology. He is currently working on analysis of the policy and media discourses around the "knowledge society" and impacts of the Internet on communication of science.

Türk, Volker is Project Coordinator at the Wuppertal Institute for Climate, Energy and Environment. He studied Land Resources Management and Environmental management and Policy in Germany and Sweden. His work currently centres on the sustainability assessment of new technologies. Volker has worked extensively on Information and Communication Technologies, Nanotechnologies, and Technology Assessment, and he was involved in various projects and activities with the European Commission, national governments, universities, business and NGOs. Co-ordinating the Nanologue project, he is member of several expert committees on the societal implications and governance of new technologies.

Wehle, Christiane studied political science at the University of Bonn. Since 2002 she has been a co-worker at the EU-Bureau of the Federal Ministry of Education and Research in Germany where she is responsible for the National Contact Points ERA-Net, Scientific Support to Policies and NEST (New and Emerging Science and Technologies) of the 6th framework program. She coordinated two EU-projects, one of which aimed to promote the dialogue with the research community on emerging scientific and technological developments (NEST-IDEA). Interdisciplinarity as well as the framework for the funding of interdisciplinary projects were among the main issues the project dealt with.

Zhu, He is project manager of Children and Youth Science Center of China Association for Science and Technology. She graduated from Peking University with a Master Degree in Law. She has been working as the chief coordinator in the research programme for preparing the National Scheme of Scientific Literacy for all Chinese since 2003. The Scheme is aimed to enhance scientific and technological literacy of all Chinese people and the Outlines of the Scheme has been formally issued as a policy document by the State Council of China in February 2006.

CHAPTER 1

WHY COMMUNICATING EUROPEAN RESEARCH?

Introduction

MICHEL CLAESSENS

Information and Communication Unit, Directorate-General for Research, European Commission, 200 rue de la Loi, 1049 Brussels, Belgium, Tel +32-2-2959971, E-mail michel.claessens@ec.europa.eu

The 'Communicating European Research 2005' (CER 2005) conference took place on 14 and 15 November 2005 in Brussels. It was organised by the European Commission (Directorate-General for Research) with the aim of providing participants in European Union's funded research projects with information on good practices and professional services in science and technology communication.

The conference was attended by over 2 100 participants, including project co-ordinators, journalists and other communication professionals, as well as press officers and representatives from research organisations.

The event focused on the manifold aspects of science communication and provided a forum and meeting place for scientists, communication professionals and journalists. Through plenary and parallel sessions, a huge exhibition and press briefings, participants promoted mutual understanding of their respective roles, shared best practice and defined strategies to improve communication, outreach and dissemination of research results to the public and the press at a European level.

The present book includes summaries of sessions that took place during the conference. The various articles cover the main aspects of science and technology communication today. They address interesting and topical issues and questions such as:

- Is science journalism necessary *at all*?
- Should communication be one of the basic skills of scientists, such as thinking, testing and experimenting?
- Should scientific literacy be a, if not the, first target of the school science curriculum?
- Do we need *science critics*?

M. Claessens (ed.), Communicating European Research 2005, 1–2.

- Does the coverage of science in the media reflect the choice of the editors or the public's interest?
- How does one promote science on local, commercial or entertainment radio?
- Why would we need standards in science communication?
- How does one remove the obstacles which impair communication *among* researchers and *between* researchers.

The different chapters in this book cast light on these issues and many others related to science communication today, and some of them provide original insight, in particular on the relationships between scientists and journalists, on Internet communication, etc. Several articles recall the fact that science communication is an integral component of the production fabric of scientific knowledge: communication *of* science cannot be isolated from communication *in* science (see, for example, Roland and Roffi *et al*). Other articles touch upon the media and supports of science communication, and show in particular that radio is effective in putting a local issue in a global context.

In parallel to the sessions, over 200 journalists attended the media briefings where latest research results and current scientific activities were presented. In the exhibition hall, 250 stands featured selected research initiatives, as well as the communication strategies of research organisations. In addition, dissemination networks, media associations, relevant publications and editorials, together with companies and service providers in the field of information and communication, presented their products and services.

CER 2005 highlights the fact that communicating research is a European Commission priority in the framework of its research programmes. The Commission wishes to draw the attention of participants in the European Union's funded projects to the fact that they can no longer ignore the 'public communication' dimension of their activity.

According to a survey published in June 2006 by The Royal Society, 70% of UK scientists believe that "Funders of scientific research should help scientists to communicate with the non-specialist public" and 46% of them do not "feel well equipped to engage with the non-specialist public"[1].

Dissemination of results is a contractual obligation for participants in research initiatives supported under the European Union's Sixth Framework Programme 2002–2006. The specific aim of this provision is to promote knowledge sharing, greater public awareness, transparency and education. Communication is key in a knowledge-based society.

The European Commission's Directorate-General for Research is heavily involved in communicating the results of European Union's funded research to the media and also to the general public. Support and help are provided to assist project coordinators and team leaders to generate an effective flow of information and publicity about the objectives and results of their work, the contributions made to

[1] The Royal Society, *Science communication Survey of factors affecting science communication by scientists and engineers, June 2006.* http//www.royalsoc.ac.uk/downloaddoc.asp?id=3074

European knowledge and scientific excellence, the value of collaboration on a Europe-wide scale, and the benefits to European Union's citizens in general.

Together with guidelines and information publicly available, the CER 2005 conference was one of the activities undertaken by the Commission to help researchers design and implement their CER activities.

By encouraging participants in European Union's funded projects to develop 'public communication' activities, the European Commission highlights the fact that they are also in an excellent position to improve the image of science and technology among a broad public.

CHAPTER 2

THINKING SCIENCE, TALKING SCIENCE

NICOLAS CHEVASSUS-AU-LOUIS

Independent scientific journalist, Les Cévennes N7, 949 avenue Louis Ravas, 34080 Montpellier, France, Tel +33-4-67541530, Email nchevassus@wanadoo.fr

Abstract: The aim of this article is to demonstrate that the main schools of thought in the field of scientific communication must be viewed within the context of a much broader debate, encompassing ideas anchored both in philosophy and sociology, on the status of scientific knowledge and on its specific characteristics in comparison to other forms of knowledge. After setting out the context in which scientific communication takes place today, using the terminology of communication theory, this article will show that the gradual shift from the notion of a 'deficit model', i.e. one which aims to instruct, to the concept of dialogue between the world of science and the general public can be seen as the consequence of an increasingly probing analysis of the true autonomy of science

Keywords: Science communication, Theory, Science

"How can such an ancient practice as science provide us with truths that are valid across all eras, independent of history and free from all links to space and time; in other words truths that are both universal and eternal?"

Pierre Bourdieu, in Science de la Science et Réflexivité,
Paris, Raison d'Agir, 2001

At the end of the 1940s, Claude Shannon and Warren Weaver suggested that all communication activities could be considered as a process involving six elements: a *transmitter* sending a *message* encoded in a *language* to a *receiver* via a *communication channel*. Lastly, *noise* is the term used in this Shannon-Weaver model to refer to all phenomena which interfere with the message between its transmission and reception. This conceptual framework is valuable in that in enables all communication to be defined in concrete terms, whether the parties involved are people or not. This framework is particularly relevant when endeavouring to describe the changes scientific communication has undergone over the last 30 years.

M. Claessens (ed.), Communicating European Research 2005, 3–11.

1. INCREASINGLY COMPLEX COMMUNICATION PROCESSES

Here we are not going to look into the issue of the messages sent. This is not relevant to our study. The issue of language diversification will be considered later. To begin with, therefore, the focus will be on the two ends of the process - the transmitter and the receiver – and the channels linking them. This section will expand on the following four ideas:

Multiple transmitters: the scientific communications field is no longer limited to a small group of prestigious scientific outlets, but now also encompasses laboratories, institutes, journals, sponsors, etc.

The restructuring of channels which is gradually favouring a multimedia approach in place of the traditional written word and print media

The fragmentation of receivers: the 'general public', presumed not to have any specialist knowledge, is no longer the sole receiver; rather other "niche" target groups, such as specialists, interest groups, funding institutions and state authorities, have emerged

The emergence of new sources of noise, such as . . .

In order to measure the evolution that has taken place, it is necessary to look back approximately 30 years to a time when scientific information intended for the general public was produced essentially by researchers themselves and, in particular, by *specific* researchers used by the media – at that time essentially the state-controlled press and audio-visual media – as uncontested sources. That era is now over for at least two reasons: i) research has become more specialised and is increasingly carried out collectively (consider, for example, the hundreds of co-authors who have contributed to publications on high energy physics and genome sequencing), with the number of researchers who fully understand and can explain all of the work done in one domain dwindling, in some cases to zero; ii) the crisis of confidence in certain aspects of science and technology underlined, once again, by the 2005 Eurobarometer surveys of the attitudes of European citizens towards S&T, and which manifests itself as a reluctance to trust one single scientist, no matter how prestigious. I will return to this point later.

Faced with this crisis of legitimacy, which in some European countries came hand in hand with a financing crisis, the world of science was called upon to justify the funding it received. This, in turn, led to a multiplication in the number of transmitters of scientific information. These were no longer limited to researchers disseminating their findings, but began to include the institutes that employ them, the journals that publish them (often business ventures competing to publish the most groundbreaking articles) and the organisations that fund them.

This multiplication coincided with a mushrooming of the number of receivers. A shift took place, moving away from a hierarchical model within which a small group of scientists were responsible for communication with a very broad and supposedly homogeneous audience and towards more targeted communication designed for specific subsections of the public. This phenomenon was particularly pronounced in biomedicine as patient organisations gained in stature. A system emerged within which all institutions began to communicate with one another, oscillating continuously between the role of transmitter and receiver. Charities publicising the progress made by teams they are funding are communicating both with their own donors and with rival foundations. Similarly, universities publishing the research carried out by their laboratories are conveying a message both to potential students they are keen to attract and to competing universities.

Despite this proliferation of information and exchanges, the amount of space devoted to science and technology news by the European mass media has remained stagnant and, in some cases, has even decreased. Scientific columns in the mainstream press have vanished into thin air and science is often dealt with in terms of the problems it is suspected of causing and not the results it achieves. This very same logic is applied by television broadcasters and, according to Eurobarometer, television – as opposed to either the press or the internet – is the medium of choice of European citizens looking for information on S&T issues.

On the other hand, scientific documentaries are flourishing, in particular in the United Kingdom, Germany and the Nordic countries. This reflects the fragmentation of the public into niche markets, catered to by specialist channels and journals. Indeed, with the exception of the *New Scientist*, which sells 140 000 copies each week, the majority of popular European scientific journals are facing a gradual erosion of their often greying readership. This decline is attributable in part to the great success of the internet. Why purchase a magazine for the purpose of reading summary articles if the very same information is available free of charge on the web? But this is also partly due to the success of specialist magazines which look not at scientific news as a whole, but deal with such areas as astronomy, cosmology, the environment or biotechnology, placing these issues within the reach not only of professionals but also the enthusiastic amateur.

In order to complete this overview of the context within which the major trends in scientific communication have taken place, it is important to mention the shockwaves caused by the severe crises of the 1990s (genetically modified organisms, prion diseases, dioxin poisoning, etc. with respect to foodstuffs and SRAS and avian flu in the domain of health). These crises transformed scientific communication, which took on a new guise as an indicator of risk. There is no doubt that a large share of the concerns surrounding the potential health risks of GMOs were the natural successor to the panic unleashed by mad cow disease and the infamous breach in the species barrier. Transgenics and the possibility of inserting genes from other species into plant genomes are perhaps scientifically quite different issues to that of feeding cattle with animal meal, but in the minds of the general public, there is evidently an overlap between these two essentially concurrent issues. This is an excellent example, using the terms of communication theory, of the *noise* that can contribute to distorting a message between emission and reception.

Thus far we have not discussed the matter of the increasing diversity of language used for scientific communication. And yet this has been a major feature of the developments that have taken place over the last 30 years. The typical interviews of the 1960s in which scientists, whose names were invariably preceded by lists of titles which served to stress further their authority (Professor X, the renowned Doctor Y, etc.), bestowed their wisdom on humble journalists are viewed with nostalgic mirth today. The images which accompanied these reports, featuring the researcher dressed in a lab coat, looking pensive and surrounded by symbolic objects, such as microscopes and various measuring devices, are characteristic of the portraits of great scientific thinkers produced during the classical era. It is with a touch of irony that we read books in which renowned scientists trace back their own careers along an almost linear and purely rational time scale, culminating in their greatest discovery. These books were often used by scientists to publicise their opinions on the most varied of topics, often far removed from their actual area of expertise. And again it is with some astonishment that we look back over television programmes designed to look like conferences set in lecture halls, during which the camera would rest motionless on the researcher as they learnedly set out their indisputable knowledge.

Articles written by a sole contributor, books and conferences, which unfailingly used the tone and register of the academic report, were the three vehicles used to convey scientific information until some 30 years ago. And while these genres have not disappeared and, in some instances, have acquired renewed support, they now form only part of the linguistic range used to communicate science. Science itself has invented a host of new methods, all of which advocate, above all, interaction with the audience. Consider, in no particular order, public debates (science cafés, interactive fora, etc.), the use of images (exhibitions, television documentaries, etc.), storytelling (fictional documentaries, science fiction, etc.) and even entertainment (scientific theatre, etc.). But these different approaches do have one thing in common: they turn their back on more traditional strategies of popularisation. They no longer endeavour to explain scientific concepts in overly *simplistic* terms; rather they aim to approach science as one might look at literature, cinema or sport, i.e. as a further element in the fabric of our culture. The aim here is to show how this diversification in the languages used to communicate science can be seen as a consequence of the shift of ideas which led the world to rethink whether science is truly separate from all other areas of human activity.

2. RECONSIDERING THE AUTONOMY OF SCIENCE

René Descartes, Galileo Galilei and Isaac Newton are widely considered to be the founding fathers of contemporary science. But in order to fully understand how modern science emerged and became a factor in producing universal knowledge, it is necessary to look beyond the biographies of these renowned thinkers, and even beyond their less famed peers. At the end of the 1930s, American sociologist Robert Merton first considered the emergence of science as a virtually autonomous social subsystem, quite distinct from the economy, religion and politics. According to Merton,[1] this development resulted from a boom during the 17th century in the intellectual

[1] Robert Merton, *The sociology of science. Theoretical and Empirical Investigations*, Chicago, University of Chicago Press, 1973

professions, in the broader sense of that word: encompassing the arts, law, medicine, technology and administration. This momentum was itself a response to the English thirst for expansion and came hand in hand with greater institutionalisation of the professions themselves. In 1662, the *London Royal Society* was created as a platform for exchange and debate within a specific procedural framework. As such, it was a key contributor to modern science and similar steps were rapidly taken on the continent, with the creation of academies of science in France in 1666, in Prussia in 1700, in Russia in 1725 and in Sweden in 1739. These institutions not only helped organise and structure international scientific exchanges, they also took on the role of regulating communication in science through supervision and evaluation. These academies could decide to accept or reject manuscripts, based on sets of criteria which are still used today. Rather than conducting their experiments publicly before a group of peers, scientists began to draft a methodology section in all of their reports (comparable to the 'materials and methods' section of scientific publications) which enabled the experiment described to be reproduced. Through a system of awards and remuneration, the academies enhanced the professionalism of scientific activity and established an inherent hierarchy within the scientific world. According to Merton, science largely has these institutions to thank for ensuring its status as a distinct social subsystem that was relatively independent of the powers of the time, which in turn enabled the scientific community to develop in accordance with its four fundamental guiding principles:

Universality, which assumes that a truth can only be established after applying a series of pre-determined rational criteria and can in no case be judged according to the merits, whether personal or social, of those who produce knowledge.

Communalism, according to which the findings of science as a whole are public property which must be freely circulated between individuals, laboratories and nations.

Disinterestedness, not in the sense of altruism, but anchored in the belief that scientists are driven solely by the quest for truth.

Organised scepticism, or more specifically the fact that researchers abstain from all non-scientific comment when assessing the work of their colleagues.

It seemed important to mention Merton at this juncture, not only because he is one of the founders of the sociology of science, but also because to date no one has succeeded in expressing these values of universalism, communalism, disinterestedness and organised scepticism better than Merton. These values are assumed to be the very basis for the autonomy of science and to impart upon the discourse of researchers an unequalled aura. When viewed in conjunction with the fundamentally positivistic attitude which developed in particular in the 'Latin countries' during the nineteenth century, this Mertonian branch of sociology gives us an understanding of the different methods of scientific communication which drew their inspiration from academia and which strove to enlighten a supposedly ignorant and uneducated public.

It was not until the 1970s that this model of 'public understanding' first began to flounder. At that time, it became clear that this type of scientific communication provided an inadequate response to growing concern surrounding the consequences of certain technological developments in science. Despite the best televised efforts of the most renowned physicians to promote support for nuclear energy and provide reassurance that it was safe, public opinion remained distrustful. This crisis of legitimacy was exacerbated further by the accident at Chernobyl in 1986. As pointed out by two French sociologists, "We are undoubtedly overestimating the cultural effectiveness of such communication systems as the mass media (radio, television, general press), which are able to reach out to a very vast audience thanks to the technological revolution that has swept the world of communication. Such systems were set up to listen and respond. They boast considerable social authority, but in fact reproduce none of the characteristics that ensure the effectiveness of teaching. While the influence of the mass media on typical behaviour in terms of consumption, elocution and conversation cannot be denied, we still have a long way to go to demonstrate their ability to transform, let alone revolutionise, logical customs, ethical attitudes and technical know-how".[2] In other words, the scientific discourse of the university framework, based on a one-on-one link between transmitter and receiver, is poorly suited for use by the media. In addition to the failure of the model of public understanding to respond to a series of concerns being expressed through the criticism of science and technology, its very theoretical basis was called into question by what is referred to as the "new sociology of science".

3. THE SHOCKWAVE OF THE "NEW SOCIOLOGY OF SCIENCE"

In short, Merton and his disciples supported three theories: i) scientific knowledge isdifferent to all other forms of knowledge; ii) scientific institutions must be independent from all other forms of social activity; iii) sociological research must strive only to elucidate the social conditions within which scientific activity is carried out, while the content of science is the sole remit of scientific philosophy.

Indeed, it was none other than the scientific philosophers who suggested that the third of Merton's theories might need to be rethought. This, in turn, weakened his first two precepts. From the 1950s onwards, the hypothesis according to which the cognitive content of scientific publications should not be critically analysed was attacked fiercely both in terms of logic (by authors such as Karl Popper and Willard Van Orman Quine) and history (Thomas Kuhn, echoing Gaston Bachelard and Alexandre Koyré). In his famous book *The Structure of Scientific Revolutions*,[3] Thomas Kuhn demonstrates that science does not advance by accumulation, rather through a series of ruptures, and that these ruptures are clearly linked to both economic and political history. The transition from the Ptolemaic to the Copernican view of the world is a very good example of this type of paradigm shift. During the Renaissance, the reference framework for science was no longer comprised of ancient and sacred texts. Rather, experiment and observation gained the upper hand.

---

[2] Claude Grignon and Jean-Claude Passeron, *Le savant et le populaire : misérabilisme et populisme en sociologie et en littérature*, Paris, Le Seuil/Gallimard, 1989

[3] Thomas Kuhn, *The Structure of Scientific Revolutions*, Chicago, Chicago University Press, 1962

And it was against this social backdrop that Copernicus came up with his heliocentric approach, without actually making any new observations. Kuhn, without commenting on the respective merits of the statements he chooses to analyse (essentially in the field of physics), demonstrates that scientific activity takes place within a social system that controls its members and their activities, but that that system also provides orientation and cognitive representations which influence the development of scientific knowledge.

In so doing, Kuhn opened up a new philosophical breach which marked a turning point in how science is viewed.[4] Two trends in sociological research in science emerged. The Edinburgh school was characterised by authors such as Barry Barnes[5] and David Bloor.[6] It considered the content of science in the same sociological terms as any other kind of knowledge. To this end, it adopted four methodological principles: i) causality, i.e. the aim of research must be to elucidate the conditions in which scientific statements are made; ii) impartiality (which was already touched on by Kuhn), i.e. the notion that the validity of a statement is not taken into account; iii) symmetry, or the obligation to explain through the same types of causes both true and false beliefs (rather than explaining the former through logic and reason and the latter in terms of social factors); iv) lastly, reflexivity, according to which the principles uncovered must also be applied to sociology itself.

The second trend to emerge from the breach opened up by Kuhn focused on the ethnography of science and suggested that science should be analysed as it is carried out from the ethnologist's point of view. Authors such as Karin Knorr-Cetina,[7] Bruno Latour and Steve Woolgar[8] undertook a host of case studies looking at researchers, controversial issues, laboratories, texts, and so on. This approach rejects the claim by scientists that they are interested solely in pure fact and they in no way mediate between the 'reality' they are studying and the findings they produce.

The programme of the Edinburgh school and the plethora of ethnological monographs of science in motion reflect, either directly or indirectly, a constructivist trend in thought according to which facts neither pre-exist nor are independent from the party which has the knowledge. As asserted by Karin Knorr-Certina, "the objects of science are not only fabricated in laboratories, they are also built in an inseparably symbolic and political manner via literary techniques of persuasion, such as those found in scientific articles, via political strategies through which scientists aim to form alliances and harness resources, and via a selection process which constructs scientific facts from the inside out". Here, science is seen as ammunition, used by

---

[4] Terry Shin and Pascal Ragouet, *Controverses sur la science. Pour une sociologie transversaliste de l'activité scientifique,* Paris, Raisons d'Agir, 2005

[5] Barry Barnes, *Interests and the Growth of Knowledge*, London, Routledge and Kegan Paul, 1977

[6] David Bloor, *Knowledge and Social Imagery*, London, Routledge and Kegan Paul, 1976

[7] Karin Knorr-Certina, *The Manufacture of Knowledge. An Essay on the Constructivist and Contextual Nature of Science,* London, Pergamon Press, 1981

[8] Bruno Latour and Steve Woolgar, *Laboratory Life: The Social Construction of Scientific Facts*, Los Angeles/London, Sage, 1979

scientists in the combat for acceptance of their findings by their peers and to convince institutions of the value of the research undertaken and the need to pursue it further.

Here the second of Merton's theories, which relates to the autonomy of scientific institutions, comes under attack. In 1994, such criticism became much more systematic in the wake of the publication of *The New Production of Knowledge*[9] by five authors who were all concerned with the "new sociology of science". This document showed that the relationship between science, industry and society had undergone a profound reshuffle during the second half of the 20th century. Up to the 1950s, there had been a distinguishable split between society, on the one hand, and academia, on the other, (independent universities, scientific disciplines which evolved according to their own system of logic and the possibility for scientists to decide what was science and what was truth). But as the position of the universities weakened, the frontiers between the scientific disciplines became smudged, scientists began working for the state and for industry and a measure of control began to be exerted on the orientation and content of research programmes, and a new rapport emerged. This relationship is characterised by a cross-disciplinary approach, greater mobility of ad hoc groups of experts chiselling out solutions to urgent problems, and priority given to economic and social factors when deciding what research should be undertaken.

In 2001, three of the authors of *The New Production of Knowledge* – Helga Nowotny, Michael Gibbons and Peter Scott – developed these theories[10] and drew some very valuable conclusions and lessons in terms of what scientific communication should look like. They postulated that the distinction between nature and culture and between science and society, which had been the basis of modern science, was no longer valid. In the post-modern context of today the public is heterogeneous and comes into contact with the producers of knowledge within the framework of an "agora" which aims to develop "inverse communication" between these two groups. This theory sets great store by the social relevance of knowledge and lends its theoretical basis to the new linguistic range of scientific communication set out above and which insist on "engagement with the public": citizens' conferences, co-decision, interactive forums, audience participation, etc.

But despite its dominance within the human sciences, this "new sociology of science" has also met with opposition. The *Sokal* case of 1996 is a striking reminder of this. Let us call to mind the intellectual context of northern America, which was at that time characterised by the American *cultural studies* movement, which insisted that minority groups (women, ethnic and sexual minorities, etc.) were the victims of modern science. Taken to its logical conclusion, this theory leads us to ascertain that all cultures produce their own truths and that these may only be assessed within their

[9] Michael Gibbons, Camille Limoges, Helga Nowotny, Simon Schwartzman, Peter Scott and Martin Trow, *The New Production of Knowledge: the Dynamics of Science and Research in Contemporary Societies*, London, Sage, 1994

[10] Helga Nowotny, Michael Gibbons et Peter Scott, *Re-thinking Science: Knowledge and the Public in an Age of Uncertainty,* London, Polity Press and Blackwell Pub, 2001

own specific cultural framework. This is a radically relativistic stance which places all knowledge on an equal footing. It was in opposition to this shift that Sokal organised his infamous hoax. But the affair very rapidly took on a whole new dimension. In his pastiche, he referred to several authors specialised in the "new sociology of science", including Bruno Latour, implicitly accusing the latter of creating the very intellectual matrix through which cultural relativism prospered, a concept which he himself was determined to denounce. One might think he was in danger of throwing the baby out with the bathwater. The intense debate which arose in the wake of the Sokal affair, in any event, succeeded in drawing attention to some critical sticking points in the "new sociology of science".[11] This debate is still ongoing. But today it is evident that it is this conflictual coexistence of diverging visions of the sociology of science which forms the basis for the varied approaches to scientific communication and with that the tremendous momentum of this field of study, testified to by a host of examples set out in this work.

[11] See in particular that edition of *Actes de la Recherche en Sciences Sociales* which focuses on science (Paris, Le Seuil), March 2002

CHAPTER 3

LET'S MAKE SCIENCE THE NEXT HEADLINE

Opening speech

JANEZ POTOČNIK

European Commissioner for science and research, 200 rue de la Loi, 1049 Brussels, Belgium, Tel +32-2-2988670, Email research@cec.eu.int

Abstract: Communicating research and engaging with the public is more than a priority. It is an obligation. Not only because research and science constitute one of society's driving forces to achieve progress, but even more so, because research has a daily impact on all European citizens as a result of the choices we make and the policies we design. Research has the potential to address and answer society's pressing needs and concerns. It harbours answers to many of the questions posed everyday by our citizens and it is our collective responsibility to ensure that it serves as a vehicle for reaching out to them

Keywords: Science communication, European research

The objective of this conference is to explore how and why science needs to reach out and touch a wider audience. It aims to show that communication is a need and a must for science.

It is quite an achievement to bring together close to 3000 scientists, journalists and policy makers under the same roof, all of whom face the same challenge.

This challenge is twofold: on the one hand we need to improve the ways in which we communicate research and on the other hand, we need to improve the image of science in society.

Let me try to put this in another way: It is clear that research alone cannot solve the many challenges that our society faces today. But it is also clear to me that in order to tackle these challenges, research is very much part of the equation.

Hence one of the main mottos of this conference: "No solution without research".

Research has the potential to address and answer society's pressing needs and concerns. Research harbours answers to many of the questions posed everyday by

M. Claessens (ed.), Communicating European Research 2005, 13–17.

our citizens. But we need to think of how we communicate this and how we identify the key messages that stem from it.

We have to rethink our strategy when faced with today's challenges. On the one hand, we stay committed to the Lisbon objectives and the ever growing requirements of a knowledge society. On the other hand, we need to confront the increasing indifference of young people to science studies, and indeed to the whole European project, and the growing gap between those who have access to knowledge and those who do not.

Therefore, communicating research and engaging with the public is more than a priority. It is an obligation. Not only because research and science constitute one of society's driving forces to achieve progress, but even more so, because research has a daily impact on all European citizens as a result of the choices we make and the policies we design.

According to the latest Eurobarometer carried out earlier this year, scientists in universities are the most appreciated amongst the different actors involved in science and research. 88% of EU citizens think that scientists have a positive impact on society.

Having said that, European Union citizens in their overwhelming majority believe that scientists put too little effort into informing the public about their work.

1. WHY COMMUNICATING RESEARCH IS AN OBLIGATION

This European Commission has recognised better communication as a policy on its own merits and has earmarked it as one of its principal objectives.

An "Action Plan to Improve Communicating Europe" was launched under the guidance of Commissioner Wallström in 2005. This Action Plan is based on three principles. First, listening to EU citizens and taking their views into account. Second, communicating how EU policies affect citizens' everyday life and what added value they bring. And third, connecting to audiences by conveying the key messages to them through the most suitable channels.

Researchers are an excellent example of the added value of working together over the country borders, and research and knowledge is Europe's unique selling point in the face of globalisation.

I strongly believe that European research must serve as a vehicle for reaching out to the public and it is our collective responsibility to ensure that it does so.

In reality, almost 40% of European citizens claim to hardly ever or never read articles on science in the media for two main reasons: they either do not understand the content or do not care about scientific issues.

This represents a challenge we need to seize, by listening to European citizens and addressing their concerns. We need to help scientists to make science more attractive; we need to explain science in clear and understandable terms, and we need to demonstrate the results of science, and their use and impact on the daily life of the EU citizen.

A particular aspect is how to meet the challenge of communicating the added value of European research. There is broad agreement amongst European citizens that collaborative research at EU level will become increasingly important and that it is more creative and effective than research carried out at national level.

More and more voices are calling for increased spending at EU level on research and development, echoing the Commission's original proposal for the 2007–2013 budget. There is growing recognition that innovation is an important driver of economic growth and those who fail to follow the rhythm of innovation will simply fall far behind.

I do not want this fate for the European Union. I want it to be the most dynamic and competitive knowledge-based economy in the world. This is why we have asked for the budget to be doubled for the next Framework Programme. This is why we have identified 19 different areas where Commission and Member States should act to encourage private enterprises to invest more in research and to work, invest, research and innovate in Europe.

2. THE SCIENCE AND SOCIETY DIALOGUE

That is all well and good as a political context, but what we are talking about today is connecting with European citizens through science. For our part, we intend to approach communication in a more strategic way across all research areas, in order to make the most of the potential for communication of research.

The overwhelming majority of EU citizens hardly ever discuss science. Clearly this poses a challenge for all of us who believe that science merits a position in the European public sphere.

Why is it that three out of four EU citizens believe that because of science and technology there will be more opportunities for future generations – and yet – so few express an interest in finding out more by reading articles and watching TV shows with a scientific content? What can we do to bring these two closer together?

We need to meet two objectives. First, we should enhance the public recognition of the role of researchers in society. Second, we have to encourage researchers to acquire the necessary communication skills so that they in turn can inform society about their knowledge and discoveries.

Initiatives such as 2005's "Researchers in Europe" campaign saw just that – thousands of people from universities to public authorities, research centres and associations came together with the general public all over Europe to demonstrate the excitement of a career in research.

Our Science and Society programme has been particularly successful in supporting events that popularise science, for example by supporting science festivals together with interactive exhibitions, or innovative scientific education initiatives that reach out to young children and adults. We annually award the Descartes Prizes for Scientific Excellence and Science Communication, and this provides a perfect opportunity to showcase the achievements of European researchers and those whose life work is getting people interested in science.

3. CHANGING THE WAY WE COMMUNICATE EUROPEAN RESEARCH

Communicating research must be a dynamic, on-going process. It requires effort and commitment by all actors, the media as well as the scientific community.

Most scientific journalists agree that reporting on research is a difficult task. Not only do they have to familiarise themselves with the scientific content and verify it but they also need to translate it for the public. Tight editorial deadlines will clash with the scientific practice of taking time to validate scientific outcomes. Scientists are often put off by the media's tendency to over-simplify.

But really, if we take the time to consider this, there is much that unites scientists and media professionals. Both, after all, are defined by their independence, their objectivity, their enquiring minds, and their search for truth. And both groups face new challenges.

The explosion in television channels, and especially 24-hour news channels, and the accessibility of news via internet put new demands on those working in the media in the same way that technological advances change the working life for scientists.

The profile of viewers of scientific programmes is selective. Still, the majority of these believe that popular science programmes can increase knowledge and broaden our minds. So how do we get a wider audience for the science shows? How do we get more people to read about science and to understand how it affects their lives?

Clearly, I do not have all the answers. Indeed, one of the key objectives of this conference, perhaps the main one, is to hear your views on our plans to bring about fundamental change in the way we communicate science at European level.

What are these plans, in essence?

Seven out of ten EU citizens get most of their information via the television. We need to make better use of this medium, and so we are considering investing in audiovisual co-productions which can then be used across different media outlets at national and local levels.

We should be looking to reconnect with thc immense possibilities of radio. Radio is a wonderful medium, as it accompanies people in their every day activities in a way that nothing else does.

We currently have two complimentary internet portals serving more than a quarter of a million users from the scientific community and also the more general public: Cordis, (which by the way is celebrating its 15th anniversary during this Conference) and, of course, Europa.

We are planning to create one single Science Portal, which will streamline their content and make them more user-friendly and appealing.

In order to bring the media closer to science, we are also considering a tool to support the training of journalists in European scientific issues.

We hope that these initiatives, plus others that might be developed during this conference, will increase our joint capacity to deliver and provide access to knowledge.

CONCLUDING REMARKS

I know that we can all agree on the need to improve our efforts in communicating science for all the reasons I have already outlined.

This conference, as I said before, is a unique opportunity to hear your views on our ideas for moving forward and improve them by learning from the many examples of excellent practice that are showcased over the next two days.

Last, but not least, let's come out of this conference with a sense of community and purpose. Our final goal is that the 7th Framework Programme will be inspired and guided by this need to make society more keenly aware of the relevance of science.

Let us make science the coolest subject in school, the hottest topic at the dinner table, and the exciting front page headline.

CHAPTER 4

INFORMATION AND COMMUNICATION TECHNOLOGY RESEARCH AND ITS IMPACT ON GROWTH AND JOB CREATION

VIVIANE REDING

European Commission, 200 rue de la Loi, B-1000 Brussels, Belgium, Tel +32-2-2981600, Email viviane.reding@ec.europa.eu

Abstract: Information and Communication Technologies (ICTs) play a crucial role in growth and job creation. They add value to goods and services, make business and government processes more efficient, and deliver the services that citizens and businesses need. Competing nations, such as the US and Japan, understand this and proportionately spend more on ICT research than Europe currently does. And in the face of global competition, ad increasingly complex technologies and soaring research costs, there is a growing need to step up collaboration in ICT research and innovation both at Member State and Community level, underlining the importance of Community efforts in this area. That is why Europe needs to invest more heavily in ICT research and get more out of it – one of the central pillars of the i2010 initiative, an Information Society for Growth and jobs

Keywords: Information and communications technologies, ICT research, i2010, Information Society, Growth and Jobs

More than ever before, research and innovation are high on the policy agenda in Europe, as policy makers recognise their essential role in the economy and society. This was confirmed by Heads of States and Governments in their Informal Meeting held at Hampton Court Palace in the UK at the end of October 2005.

What matters now in 2006 is to turn this political momentum into concrete action and into real commitments to invest more and better in research and innovation.

Today the race towards more knowledge, to master technology and its best use, and to develop innovative high value products and services, is global. All major established or emerging economies have realised the need to be part of this competition in order to prosper and grow.

M. Claessens (ed.), Communicating European Research 2005, 19–23.

So words and declarations no longer suffice. Europe has strong and unique assets that enable us to compete and succeed, but we need urgently to adjust quickly to the changing economic realities brought about by globalisation and the ever-faster pace of technological change.

We know that we are lagging behind our direct competitors in investing in research and development in general. What we perhaps know less well is that the largest gaps are in the technology areas that are essential to increase our growth potential and to address current and future societal needs. This is the case, in particular, in the key field of Information and Communication Technologies.

1. INFORMATION AND COMMUNICATION TECHNOLOGIES

Information and Communication Technologies (ICTs) play a crucial role in growth and job creation. ICTs add value to goods and services, make business and government processes more efficient, and deliver the services that citizens and businesses need.

The latest statistics show that the ICT sector represents over 5% of the EU GDP. This 5% drives 25% of overall growth and about 40% of our increase in productivity. Still this is less than our main competitors.

A wider and more efficient *use* of ICT throughout the economy and society is needed to help Europe strengthen its global competitive position and address the challenges ahead.

All recent analyses show that the economies that invest in research in a particular technology are those that benefit the most from this technology and its progress. Those that rely only on knowledge developed elsewhere are ultimately left behind.

An indigenous research capacity is therefore essential to be able to master and assimilate technologies and to exploit them to economic and societal advantage. This is true for all new technologies, and it is particularly true for ICT:

- where the frontiers of research are increasingly broad,
- where the pace of innovation is accelerating, and
- where people and organisations increasingly depend on ICT.

Today we are far from exploiting all the possibilities that progress in ICT can offer to people, businesses and governments. A world of opportunities is ahead of us and Europe cannot afford to miss them. The next waves of ICT development will make devices, networks and services smaller, smarter, cheaper and more reliable. ICT will provide a wider range of applications, enabling people to interact with their surroundings and with each other in totally new ways. Progress in ICT will help us monitor our health and well being at anytime and anywhere. It will enable every child and adult to get access to personalised learning whenever needed. It will bring infinite computing and data handling capacities to the desk of every engineer and scientist and will enable every business, large or small, to be connected to worldwide trading communities.

This is why, in most major economies, ICT research gets the largest share of research investment, accounting for over 30% of the total research effort. This applies to both the public and the private research fields. However, this is not the case in Europe, where ICT research represents only 18% of the total research effort.

Today, Europe invests only a third of what the USA invests in ICT research and 30% less than Japan. This is undermining Europe's capacity to compete in all sectors of the economy, to meet the demands of our societies and to innovate in all science fields. Europe should bridge this gap and we are certainly capable of doing it.

Strong and well-coordinated research investments in ICTs in the 80s in particular have enabled Europe to become a world leader in many ICT sectors. Telecommunications, embedded computing, micro- and nanoelectronics, micro-systems, "smart" integrated systems and rich audiovisual content are Europe's main industrial and technology strengths.

Currently, Europe represents around one third of the worldwide ICT market and 20% of world ICT supply. We can build on these assets to remain a key player in these strategic technologies. But for that we have to invest more and better in ICT research.

This is why the Commission has conducted research *in* ICTs and innovation *through* ICTs, the central pillar of the newly launched i2010 initiative – "A European Information Society for Growth and Employment". Europe needs urgently to put more into ICT research and to get more out of it.

2. TURNING TECHNOLOGICAL PROGRESS INTO INNOVATIVE APPLICATIONS

This conference is about exchanging best practices in science and technology communication. So let me first illustrate what I just said about ICTs and their application with a number of concrete examples.

Experience has shown that it is through pan-European co-operation and coordination that a critical mass can be created, that leadership can be developed and that European technology can become the worldwide standard.

For example, ICT-based anti-collision radars and automatic braking systems could reduce rear-end collisions by up to 75%. Lane change assistance could save 1,500 lives per annum if only a relatively small percentage of all vehicles were equipped with it. EU funds are contributing to these developments: for instance the PREVENT project that works on preventive road safety technologies. Total private-public investment in PREVENT is close to €60 million, of which the Commission is contributing half.

Another example from the transport sector: Time-Triggered Architectures is an innovative EU-funded technology that provides highly reliable electronic systems where safety issues are critical. Today these architectures are deployed, for instance, for the Airbus A380 cabin pressure system, as well as for electronic braking systems and engine controls in cars. It is also used for controlling electrical power plants and large-scale telecommunication networks. In the health sector, too, progress in ICT is providing innovative solutions. Electronic health records for example help reduce the number of fatal medical errors estimated to affect tens of thousands of people in Europe. Regional health data networks, keeping all the information exchange electronically, has shown to reduce costs considerably. In Denmark, for instance, savings from a system of electronic booking from general practitioners' offices to hospitals currently save millions of euro each year. Early EU investments in the R&D of the underlying technologies have played an important role in these results.

Today, Europe hosts world-leading manufacturers of low power and digital signal processing devices: The leading European chip manufacturers – STMicroelectronics in France and Italy, Infineon in Germany and Philips in the Netherlands – have been very successful in mastering the micro and nanoelectronics technologies. They invested billions of euros to exploit the outcome of their research. Thanks mainly to European and national research programmes, all three are now among the top-10 semiconductor manufacturers worldwide whereas there were no Europeans in the top-10 in the early 90s. In this example it is essential to stress also the impact on the creation of jobs. STMicroelectronics, for instance, employed 22.000 ten years ago – half in Europe, half elsewhere. Today, they employ 49,500 – half in Europe, half elsewhere.

More generally, clustering in high-technology hotspots has generated jobs and economic growth. For instance, the Dresden zone has created 15,000 permanent jobs at semiconductor plants and attracted more than 30 international equipment and service suppliers. The Grenoble zone has generated over 6,000 jobs directly and added associated activities that include over 20,000 total jobs linked to nanoelectronics.

Another example is in broadband data networks: The roll-out of the new generation of broadband data networks in Europe has been largely based on DWDM – an optical technology used to increase bandwidth over existing fibre optic backbones – and on low-cost ADSL access modems – a technology that allows more data to be sent over existing copper telephone lines. European equipment manufacturers have played a leading role in the development and commercialisation of these key communication technologies, which have also been supported through European level collaborative research.

3. STEPPING UP COLLABORATION IN ICT RESEARCH AND INNOVATION

Today's context of growing global competition, increasingly complex technology chains and soaring research costs point to the urgent need for Europe to increase substantially its public research effort in ICT and to pull it together so we that can shape the future and compete in the global race.

It is equally essential today to strengthen the effort at Community level. More than ever before, cooperation in ICT research is important in order to share costs and risks and to combine the key multidisciplinary skills that are needed to succeed.

Increasing and pulling together the ICT research effort in Europe both in the Member States and at Community level will exert a significant leverage effect on private investments in R&D. We see today that those emerging economies with significant public investment in ICT research are those that attract the most private investment in this field.

A strategic approach to ICT research and innovation has served Europe well. Experience has shown that in areas where a focused research effort was undertaken at European level, important successes were achieved. This is what we are trying to replicate today, not only with direct support to transnational collaborative R&D projects, but also with the promotion of the concept of European Technology Platforms.

4. ICT IN FP7 AND THE CIP

With the Commission's proposal to increase the support to ICT research in the Seventh Research Framework Programme 2007-2013 (FP7) and to ICT take-up and use in the Competitiveness and Innovation Programme (CIP), we have tried to show the way.

However, when I looked at the debate on the EU Financial Perspectives for 2007–13 that took place at the end of 2005, I became very concerned. Our position in this debate was clear:

- research and innovation is a priority for Europe; it is at the core of the renewed Lisbon strategy for growth and jobs.
- Increased support for co-operation and collaboration in ICT research is crucial; it has shown its worth in the past and it is becoming increasingly important in the new global context.

CONCLUSION

We need now to invest more and more wisely in research. We need also to keep the right balance in our approach: Europe should indeed play a key role in pushing science and technology beyond current frontiers. But we should also make sure that research enables us to address current and future societal and economic challenges.

It is important to keep in mind that "knowledge for knowledge's sake" is often necessary but is far from being sufficient. We need to ensure that a solid and continuous chain of knowledge exists, from fundamental and curiosity-driven research through to industrial research and innovation.

The launch of the European Research Council, ERC, in FP7 will help us support fundamental research through individual grants at EU level. This has to be combined with a strengthening of co-operative research. This is where industry-academia collaboration across Europe is helping us boost innovation and competitiveness in all sectors. For the ICT field and hence for the whole economy, this collaboration is becoming essential.

CHAPTER 5

THE EVOLVING CONTEXT FOR SCIENCE AND SOCIETY

ALAN I. LESHNER, PH.D.

American Association for the Advancement of Science, and Executive Publisher, Science, 1200 New York Ave., NW, Washington, DC 20005, USA, Tel +1-202–3266639, Email aleshner@aaas.org

Abstract: The relationship between science and society is undergoing substantial change, some of it quite negative in character. Some tension arises from different ways of evaluating the relative risks and benefits of new findings and technologies. Other sources of tension have to do with the way some scientific findings are encroaching upon issues of core human values. One consequence is increased efforts by the public and policymakers to unduly influence both the course and conduct of science. The relationship can only be improved through engaging the public in a true dialogue about their concerns about scientific issues and what to do about them

Keywords: Science and society, Public engagement, Public attitudes, Communicating science

Science and technology are critical elements in every aspect of modern life. Science-based innovation is central to economic prosperity, and scientific advances have contributed enormously to improving the quality of life and health. There are virtually no issues or aspects of current society that do not have some science and technology component to them. This centrality of science and technology means that in order to live full and productive lives, citizens need both familiarity and comfort with science and scientific issues.

Progress in science itself, in turn, is dependent on a receptive, even appreciative society that must not only provide support for the conduct of science but must also be equipped to use the products of science in the most effective ways. That makes the societal context for science critically important, both to its success and to its usefulness to the people it is intended to benefit.

The current context for science is changing. There is a palpable increase in tension between science and members of the lay public. This tension is resulting in large part from an encroachment of science onto issues of human values, as discussed below.

M. Claessens (ed.), Communicating European Research 2005, 25–30.

The solution or resolution of that tension will depend on the willingness of the scientific enterprise to enter into a genuine dialogue with members of the public about their interests and their concerns, and what to do about them.

1. PUBLIC ATTITUDES

To be clear, whenever the general public is surveyed, we see generally positive attitudes; most people (70–90% in the United States) believe that the benefits of science outweigh the risks. However, in many places that overall view of science has deteriorated over the last decade. For example, in Europe, a 2005 survey[1] found that only 52% of people thought the benefits of science outweighed the risks, whereas in 1992, 61% of people felt this way. Moreover, even though the numbers in the United States sound better, many Americans really do not understand what is and is not science. For example, 60% of Americans believe in extrasensory perception and 41% believe astrology is somewhat scientific.[2]

2. THE NEW VALUES DIMENSION

Some of this change in the societal context derives from a new dimension emerging – or re-emerging – in the science-society relationship. Whereas historically, science and technology have been evaluated primarily on the basis of their relative costs or risks versus their benefits, we now are seeing issues of how science relates to human values being brought into the equation.

This values dimension is not really new, of course, since throughout history there have been many examples of the incursion of value issues – like whether the earth revolves around the sun or, more recently, whether scientists should work on nuclear weapons. However, right now there seem to be many more values-related issues on the agenda, and the effect on the science-society relationship is significant. Examples of current values-related issues include cloning and stem cells, whether to study such "personal topics" as sexual behaviour or the genetics of individual traits, and, of course, whether to teach "intelligent design" (ID) as an alternative to evolution in science classrooms.

The intelligent design/evolution issue is a clear example where scientific understanding can conflict with core values or religious beliefs. ID is a belief concept that says that although humans came to be as they are gradually over time and may have come from earlier life forms, there has been a supernatural intelligent designer who crafted the entire process and has been guiding it in explicit detail. ID proponents claim this is an alternative scientific view to evolution and that it should be taught as such in science classrooms. Scientists, on the other hand, object, pointing out that there is no science base to ID and, in fact, the very notion of an intelligent designer is not a scientific question. It cannot be tested scientifically. Scientists argue that ID

[1] European Commission, *Europeans, Science and Technology, Eurobarometer Survey*, 2005, EUR 21722.
[2] National Science Board, *Science and Engineering Indicators, 2004*.

is fundamentally a religious concept, should not be confused with science, and should be kept out of the science classroom.

Without spending too much time discussing intelligent design, the point here is that the overlay of values is having serious consequences for the entire science-society relationship, and in two fundamental ways. First, members of society, including many policymakers, are trying to regulate or control science in ways they have not in many years. Second, it is contributing greatly to the tension referred to above. This overlay of values – or the conflict between science and some values – is creating a growing divide between science and the rest of society.

3. OTHER CAUSES OF THE DIVIDE

This abutment of science against issues of human values is only one cause of the science-society divide. Another is a fundamental misunderstanding, alluded to above, of the very nature of science and its core concepts. As one example, laypersons and scientists use the term "theory" quite differently, and this causes great confusion. To laypersons, a theory is an educated guess. To a scientist, however, an educated guess is more often termed a hypothesis, and one only gets to call something a theory after much empirical evidence has been collected, many hypotheses tested, and a scientific base for the theory has been generated. In science, a theory typically is the end product of a tremendous amount of prior scientific work, not the beginning or a precursor of scientific work. Only after the scientific data base has been generated can one claim a theory as a way of organizing concepts or explaining an important aspect of the natural world.

Another major cause of the growing divide between science and the rest of society is the presumption by scientists that people's objections to what science is doing or showing arises only from misunderstanding on the part of the public. Scientists frequently attribute public disagreement simply to public ignorance. That is often wrong. In many cases, the public does understand what science is doing or showing, but people simply do not like it – or it conflicts with their core values and therefore they object. Research on embryonic stem cells and/or cloning is an excellent example. Studies show that many people who object to embryonic stem cell research understand full well what it is about. However, it conflicts with their values or beliefs about the nature and origins of life, and therefore they do not want that line of research pursued.

This is not to imply that educating people is not useful. As I suggested before, there are many instances where people do misunderstand science or have no idea what we are talking about. In those cases, education is essential. However, the message here is that that we cannot simply educate our way out of the growing divide between science and society!

4. ENGAGE THE PUBLIC

The most common response from the scientific community to the public's increasing desire to influence the course of science – whether by increasing regulations or by overly directing the research agenda – is to rise up in protest. We say "leave the scientific agenda to the scientists and let us police our own activities".

That approach, however, is unrealistic and unproductive. Although it is true that the scientific enterprise has done an excellent job both policing itself and setting its own course, that argument does not work with either the public or policymakers. To them, "leave us alone" is not a satisfactory response; they do not believe we are able to police ourselves sufficiently well. Moreover, they argue, it is clearly our independence that has brought us to the point of either encroaching on values issues or revealing facts they would rather not know.

Many people, myself included, believe the only real hope for reducing the tension in the science-society relationship is to engage in a much more open and full dialogue with members of the public. We need to move from our traditional educational stance that looks much like a paternalistic monologue, and move to a genuine conversation with the public, where we would seek common ground. I am not suggesting that we compromise core scientific values, nor that we try to engage meaningfully with radical ideological fringes, whether on the right or on the left. But there are many people "in the middle" with whom we could and should be discussing these issues in which we all have a stake.

We need to listen to the public regarding their concerns about science and technology and their concomitants: their risks, benefits and other implications. We need to listen to and allow the public to influence priorities for research. Most research is funded by tax money; therefore, taxpayers have a right to say what is the most important work to be done.

In advocating a role for the public in setting the research agenda, I am not suggesting that the public should list topics they do not want to know about or subjects we should be forbidden to study, since that would run counter to society's overarching goal for science: to reveal and explain as much as possible about the nature of the natural world. A major role for science in society is to explain the nature of the world, whether we always like the answer or not. However, there are many questions that members of the public would like to see answered that we could and should be working on. Again, our task is ultimately to be of service to society. The public should have a chance to raise some of the questions to be tackled.

5. EFFECTIVE PUBLIC ENGAGEMENT

A frequent question is: Do these dialogues work? Some do and some do not. The field of human genetics has done an outstanding job of public engagement and provides excellent examples. The US National Human Genome Research Institute (NHGRI) includes a program of Ethical, Legal and Social Implications of Research (ELSI), and this program has supported an array of projects that bring the public into discussions with scientists and policymakers. The ELSI experiences have shown the usefulness of this engagement approach to working through and responding in a genuine way to public concerns. The Pew Foundation supports a Genetics and Public Policy Center at Johns Hopkins University that has also carried out a variety of successful public engagement activities.

The UK's British Association for the Advancement of Science (BA) has as its mission "to create a positive social climate in which science, and the organizations dependent on it, advances with public consent and with the involvement and active support of non-scientists." The BA has conducted a wide range of activities, including large science festivals, small working group meetings, and a variety of exhibits and exhibitions.

Although the formats used in engagement activities can vary a great deal, there are some fundamental principles that can be derived from the successful ones. First, large audience, town meeting formats, where a panel of scientists first give fairly technical lectures, followed by open microphones where advocates on either side of the issue usurp the podium, have been uniformly un-successful. A much better approach is small group, workshop type formats where the participants from different sectors work to solve a problem together.

It also is important to select and present issues in ways that are personally meaningful for members of the public. Even global issues like evolution need to be discussed in locally meaningful terms. A concept now being touted in various circles is "going glocal" – discussing global issues but in a local context.

It is critically important in these engagement activities that the scientists stick to their roles as purveyors of facts. I have often said it is unfortunate that scientists are also people – they have personal values that they at times have difficulty keeping out of science-related discussions. But in these discussions, scientists are supposed to be the "fact people". It is critically important that scientists not mix their own values into their discussions or interpretations of scientific facts, lest they lose their credibility and ability to have an objective dialogue.

The American Association for the Advancement of Science (AAAS) is developing a Center for Public Engagement with Science, dedicated to the kinds of public dialogue discussed here. The Center will develop partnerships with diverse other organizations with public engagement activities, like science museums and centres, and will provide speakers and other forms of outreach to community clubs, lodges and religious institutions. The AAAS annual meeting recently added a "Family Science Days" component, where thousands of non-scientist citizens from the local community have come into the formal meeting and interacted with scientists, took advantage of special exhibits for children and their parents, and provided a venue for both formal and informal discussion of critical scientific issues.

CONCLUDING COMMENTS

As we look to the future, it is clear we will need to move beyond our traditional approaches to communicating science, which have focused primarily on increasing so-called public understanding of science. Do not get me wrong. We need to continue, even to expand, these activities since there is so much to teach and so many people needing to be taught.

But public education about science will not be enough. As the public has learned more about science, and as science has revealed more about issues close to people's personal interests and values, there will be a marked increase in the need to engage in new kinds of dialogues with public citizens. One set of goals, of course, will continue to include increasing public understanding, appreciation and support of science. However, there also are a set of goals related to the societal context in which science is embedded. Society supports science because they expect it to enhance the human condition. We need to make sure we are doing that in ways that are not only understandable but also acceptable to those very people we are trying to serve.

CHAPTER 6

SCIENCE COMMUNICATION ON DEMAND

DONGHONG CHENG[1] AND HE ZHU[2]

[1]*China Association for Science and Technology (CAST), 3 Fuxing Road, Beijing 100863, China, Tel +86-10-68571869, Email dhcheng@cast.org.cn*
[2]*China Association for Science and Technology (CAST), 3 Fuxing Road, Beijing 100863, China, Tel +86-10-68571869, Email zhuhe@cast.org.cn*

Abstract: By describing several science communication activities carried out by the China Association for Science and Technology (CAST) as well as resources and strategies applied by CAST, this chapter tries to express an important principle for science communication in China, that is to say activities should be designed and implemented responding to varied demands of different groups of people

Keywords: Science communication, China

China is the most populated country in the world and is still undergoing the developing process both in economy and in society. Science communicators in China are facing the challenges of many varied demands and very limited resources. Over the decades, we have learned an important principle for science communication in China. That is, science communication activities should be tailor-made and implemented responding to the varied needs of different groups of people.

1. BACKGROUND

Over the last two decades, Chinese people have seen the dramatic changes to which science and technology have greatly contributed. Thus, according to our survey, most of the public have a positive attitude towards science & technology. Their passion and concern about scientific development are not difficult to detect.

However, the requirements of different groups of people regarding science communication are very different. China is a large country in terms of geography and population. There is a big gap between the eastern and western areas of China, as well as between

M. Claessens (ed.), Communicating European Research 2005, 31–35.

urban and rural areas. Children from rural areas (a number of about 70 million persons) only go to school for an average of seven years. There are even 8 million illiterate residents, most of them living in the western part of China. These people are eager to improve their standard of living by learning more scientific knowledge and by mastering more applied techniques related to their daily lives and production. In the meantime, more than 100 million people in China have regular access to the Internet. Most of them are teenagers and "white collar workers" in China's eastern cities. Their demands as regards science communication are in cutting-edge science and technology developments and their impact on society.

As the main driving force of science communication in China, China Association for Science and Technology (CAST) has given much attention to the demand-oriented principle during its efforts on science communication.

2. ACTIVITIES MEETING DIFFERENT DEMANDS

Founded in 1958, CAST is a non-profit, non-governmental organisation of Chinese scientists and engineers, composed of 167 learned societies. It has local branches at different levels throughout China. The following are some demand-oriented science communication activities carried out by CAST.

2.1. Training Programmes and Consultation Services for Farmers

According to statistics, technical training is one of the main channels by which Chinese farmers have access to new techniques. It is also known that farmers are more likely to accept and trust face-to-face consultations and training rather than long-distance education programmes. CAST, therefore, encourages scientists to be engaged in outreach activities to provide scientific training and consultation services for farmers. At the same time, those pilot farmers who have applied agricultural techniques and therefore have gone from poverty to a more comfortable existence are encouraged to be mentors to other farmers during the training workshops. This kind of "peer-teaching" methodology was extremely welcome at our science communication programmes held in rural areas.

2.2. Hand-in-hand: Science for Teenagers

There is great demand for science education resources in China, such as equipment, qualified science teachers as well as well-designed extra-curricular scientific activities. CAST, working closely with the Ministry of Education, has been encouraging scientists and research institutes to support science education by providing popular science lectures, internships in laboratory for high-school students, guidance on teenager's science research projects, training programmes for primary and high-school science teachers, organising science fairs, setting up virtual connection between scientists and teenagers through Internet, and so on.

Every year, thousands of scientists are involved in the China Adolescents' Science and Technology Innovation Contest (CASTIC) as supervisors or judges. More than

15 million primary and secondary school students participate in this activity each year, and it has become the most influential national science fair for young people.

2.3. Science Wagons – Science Museums on the Wheel

Science museums function as centre for the communication of science to the general public, and attract millions of visitors each year. However, most of the Chinese science museums are located in Beijing, Shanghai and some other big cities. It is hard for people who live in the rural and remote areas to enjoy the exhibitions of the science museums.

Responding to the request of rural communities and schools, science wagons have been created by CAST to bring the interactive science exhibitions, popular science DVD programmes, as well as science films to people who are living in small townships and villages. In order to do this, and after a period of 5 years developing the idea, we can now see 68 science wagons on country roads throughout China.

2.4. Science Communication Responds to Emergency

Responding to the urgent need for proper scientific knowledge when the SARS epidemic broke out in 2003, CAST cooperated with the Chinese Prophylactic Society and the Chinese Medicine Society, both being members of CAST, to develop posters and brochures on prophylactic knowledge for the general public. The information is expressed in easy to-understand language styles and supported by cartoons. The digital file of these materials was put on the Internet so as to provide free download service to various social sectors, while 2 million printed posters were sent out to every village in the mainland of China with the help of China Railway Society, which is also a member of CAST.

Such an effective mechanism of science communication has been in action again during winter 2006, in the fight against avian flu.

3. RESOURCES AND STRATEGIES

As a non-profit, non-governmental organisation, how can CAST implement these demand-oriented activities? Where do the resources come from? In which way can CAST meet the local demand? The following are some answers: it is *the Law on Popularisation of Science and Technology* that provides CAST with a better macro environment and policy support; collaboration and social mobilisation makes various resources available; research helps us to make an in-depth evaluation of needs; scientists are task forces of CAST's initiatives.

3.1. Law on Popularisation of Science and Technology

On 29 June 2002, China's National People's Congress issued the *Law of the People's Republic of China on Popularisation of Science and Technology*. Provisions of this law prescribe that the China Association for Science and Technology (CAST) and its local sections constitute the main sectors of society working for popularisation of science and technology. According to this Law, CAST has the responsibility to carry out popular S&T activities, put forward suggestions to governments and support other mass organisations, enterprises and institutions in their efforts to carry out activities for Popular S&T.

The enaction and implementation of the *Law on Popularisation of Science and Technology* not only delivered an important message to the society that the State and its people support science communication as it is our nation's priority, but also lodge power with CAST for organising science communication programmes.

3.2. Collaboration and Social Mobilisation

It is realised that in order to carry out effective science communication, more extensive social participation is necessary. With the support of the Law, cooperation and social mobilisation have been set as the most important strategies.

Good partnership with governmental agencies benefits CAST's science communication initiative in many aspects, policy and finance in particular.

Cooperation with MOE (Ministry of Education) is a good example. During the new round science curriculum reform since 1998, CAST has called together scientists to participate actively in curriculum development and teacher training. As CAST's partner, MOE gives much support to our extra-curricular activities by issuing a series of policy documents, i.e. encouragement of schools and universities to participate in CAST's science fairs; pedagogic theory and practices for organising students' scientific activities should be put into the syllabus of science teacher's pre-service and in-service training, etc.

Besides the cooperation with the governmental agencies, CAST benefits widely from the partnership with enterprises and international organisations and this is crucial for improving the quality of programmes and fund-raising.

With the support of the Ministry of Finance, CAST has set up a science communication grant to provide funding for good projects so as to mobilise more participation since 2002. It is the first time that CAST fund-raises not for its own project, but for supporting others. During the last 3 years, 267 projects were granted. Grantees included scientists, medical doctors, university professors, science column writers, learned societies, as well as governmental officials.

3.3. Research Works

Effective science communication can only happen when the needs of the society are deeply understood. To reach this goal, sound research is required in order to identify different demands for, and evaluation of, the effectiveness of science communication programmes.

Against this background, CAST has been sponsoring the *Chinese National Survey for Public Understanding of Science and Attitudes towards Science & Technology* continuously sincc 1992. The objective of the survey is to get better understanding of the contemporary level of scientific literacy and the needs of science communication of Chinese adults (aged 18-69). The main contents of the survey include: public understanding of science (understanding of scientific concepts and terms, scientific process and the impact of science on society), public attitudes towards science and technology and the sources from which the Chinese public obtain their information pertaining to science and technology.

3.4. Scientists

Scientists are the task forces of all the CAST's initiatives and their most valuable resources, for those who have the passion and sense of responsibility for science communication. There are science communication committees in the organisation structure of over 60% member societies of CAST. Relying on this effective network, scientists and research institutes are mobilised to do science communication. As mentioned previously, there are local science and technology associations, as branches of CAST, in all the provinces, covering over 70% of counties. The interaction of these two "wings" of CAST, national learned societies and local branches will guarantee a good science communication programme. The needs of the local communities, schools and governmental authorities will be made clear to local branches of CAST, then to the CAST headquarters or directly to the related societies. The learned societies will organise task forces to provide a science communication service according to the various needs expressed.

4. NEW INITIATIVE

In order to effectively coordinate and integrate social resources for science communication at national level, a national long-term action plan, *National Scheme of Scientific Literacy for all Chinese*, has been formulated and will be implemented in China soon.

The idea was first raised in 1999 by the standing committee of CAST. As a long-term project, the scheme is aimed to provide better science and technology education, science and technology training and all means of science communication to enhance scientific and technological literacy of all Chinese people.

The implementation of such a large national scheme relies on the collaboration and coordination of all the parties concerned. During the last 3 years, 13 national organisations worked together to formulate the framework and outlines of the *National Scheme of Scientific Literacy for All Chinese*. Among them are the Ministry of Education, the Ministry of Science and Technology, the Chinese Academy of Sciences, the Chinese Academy of Engineering, the Chinese Academy of Social Sciences, and the National Natural Science Foundation.

It is expected that the Outlines of the Scheme will be formally issued as a policy document by the State Council early next year, and that the Scheme will be implemented under the leadership of the State Council accordingly. There is no doubt that the implementation of the Scheme will respond to the demands of (the majority of Chinese people, and will bring tremendous opportunities to science communication.

CHAPTER 7

BRINGING SCIENTISTS TO THE PEOPLE

Research Communication Training: A First Step for Science Events

CAROLYN GALE

Stanford University, Center for Teaching and Learning, 590 Escondido Mall, 4th Floor, Stanford, CA 94305, USA, Tel +1-650- 2486272, Email cgale@stanford.edu

Abstract: This chapter discusses the idea of training scientists and researchers as a direct method of communicating with the general public. It outlines some of the issues that arise when teaching scientists to translate their work to an audience outside of their sub discipline

Keywords: Training, Outreach, Science Events

1. BRINGING SCIENCE TO THE PEOPLE, BY BRINGING SCIENTISTS TO THE PEOPLE

As an invited speaker for EUSCEA (European Union Science Communication Events Association), I considered the organization's slogan of "Bringing Science to the People." I considered how my work provides a match of EUSCEA's need for scientists to effectively engage with the general public with my own need to discover additional methods to connect research communication training to the general public.

1.1. Personal and Workshop Background

My specific experience with science communication has been as the former Director of Stanford University's Research Communication Program. Our goal was to teach PhD-level researchers how to communicate the nature and significance of their work to audiences outside of their (sub)specialties. From a pilot test of 12 PhD students in 1999, the program grew to teach over 1000 PhD students, postdoctoral scholars, research staff, and faculty from over 85 disciplines at institutions in the US, Canada,

M. Claessens (ed.), Communicating European Research 2005, 37–42.

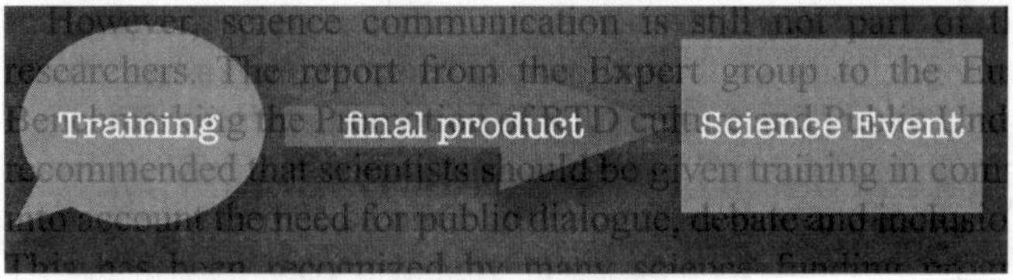

Figure 1. The result of training scientists: Can these talks directly interface to a science event?

Sweden, Norway, Denmark, South Africa, and Japan. Along with teaching, we collaborated with communication faculties across the US to research how experts learn to communicate complex information to general audiences. The following sections briefly outline what we teach in our basic workshops, along with providing some practical advice on lessons learned.

2. METHODOLOGY OF OUR WORKSHOPS

The goal of a basic research communication workshop is to enable scientists to describe their work in 750 words, understandable to a 17–18 year old student. In the US, this would be equivalent to a senior in high school, or a freshman (first year) at a college or university.

The 750 word statement is broken down into two parts: the Elevator Talk (abstract), and the Hallway Talk (body). After several rounds of peer review and instructor support, this "final product" is complete, and often ready to be converted to a 5–10 minute presentation. It is important, however, to work on the content first, which is the reason why we emphasize working on a written statement first.

One of the first things we do is dispel the myths participants bring to the workshop. First, people tend to think that translating their work for a different audience is going to be easy and not take a lot of time. After our first exercise (see section 2.1), participants realize that this is not easy at all!

2.1. The Elevator Talk

The written Elevator Talk is approximately 150 words in length, and should explain a person's field of study, exactly what he/she does, and why it is important in a broader context. This is easier said than done!

Why the name Elevator Talk? It helps to imagine that you have walked into an elevator, and someone asks, "So, what do you do?" This elevator is not in a very tall building, nor is it particularly slow. You might have 30 seconds – what would you say? We have found the analogy of an elevator to be immediately understood, regardless of where the workshop is taught.

Before a workshop starts, participants are given detailed instructions on how to develop an elevator talk, with examples and issues to consider. A few points that we try to emphasize are:

1. Include only one technical term in this talk – maximum.
2. Use an analogy to clarify a concept; or to tell a story. Look for something concrete that is easily understandable to a general audience, and use that example to transition to a more abstract theory.
3. Avoid giving nothing but background information about the research. Use action words to describe exactly what is being done.
4. Be sure to situate the work in a larger context. Why is this work important?

When all participants meet face to face, they are placed into groups of two or three. Each participant reads his or her Elevator Talk aloud, and receives guided feedback from peers (an example of this feedback would be making note of technical terms that were not understood). In this exercise, it is important to place people together from fields as diverse as possible. For example, we have learned that placing an electrical engineer, a physicist, and a mathematician in a group does not work - they understand each other's technical terms at a basic level.

2.1.1. Detecting Jargon

Here is an example Elevator Talk brought to a workshop:

"I model the diffusion of drugs through the skin, using a method called finite elements. This is to understand better how transdermal drug delivery works and also to help improve their performance. The complex structure of the skin makes this quite challenging, and I'm looking at the diffusion on both molecular scales to molecular dynamics simulations and also at the macroscopic scale through finite element calculations."

As mentioned previously, one of the first things to work on in a group setting is to identify technical terms. For an Elevator Talk, we require a maximum of one technical term (preferably none). If we highlight the first two sentences, we can see that there are four technical terms:

"I model the **diffusion** of drugs through the skin, using a method called **finite elements**. This is to understand better how **transdermal** drug delivery works and also to help improve their performance."

While talking to the author of this Elevator Talk, I put aside the writing and asked her about the work she did. Immediately, she mentioned transdermal drug delivery, but gave an example of a person who wants to stop smoking by wearing a nicotine patch. This patch delivers drugs through the skin, and that is what transdermal means. This is an example of using something concrete (a nicotine patch) that is generally understandable by much of the general public, and using that to help visualize the specific research she does (using a computer to visualize how transdermal delivery works). I suggested that she revise her Elevator Talk accordingly.

2.1.2. Less is (not always) more

Here's another example of an Elevator Talk, but we will stop at the first sentence.

"I study the effects of plant secondary compounds on herbivory."

This may be concise, but **plant secondary compounds** and **herbivory** are technical terms. This person has lost the audience after the first sentence!

When faced with the task of being concise, researchers have a tendency to "pack in" as much information as possible. This is not the point of an Elevator Talk, and we challenge participants to continue to prioritize their work. If there is only one thing the audience can take away from your written statement or presentation, what would it be?

In this case, I sat down with the student and started by asking him what he meant by plant secondary compounds. Without looking at his writing, he asked me if I had ever wondered why some plants taste bitter? Some plants do this to protect themselves from being eaten. This is what the research was about – the effects of bitter plants on animal dietary preferences.

"Have you ever wondered why certain plants taste really bitter? It's because they produce these bitter compounds to protect their leaves from being eaten. This explains why certain plants are eaten more than others."

The first sentence was rewritten into three. Yes, it may be longer, but a few more words make the introduction engaging and concise.

2.1.3. Progress, Not Perfection

Building on the last example, I continue to emphasize that researchers cannot impart everything they know about their work in one talk – or two talks, or even over the course of a typical university class. Experts tend to forget that it has taken them at least 6–10 years to gain the knowledge that they have about a subject – where a layperson might have the equivalent of six months to one year of experience in the more general topic.

I continue to stress that it is important that even if you manage to advance the knowledge of your research just a notch, you have done more than enough for one

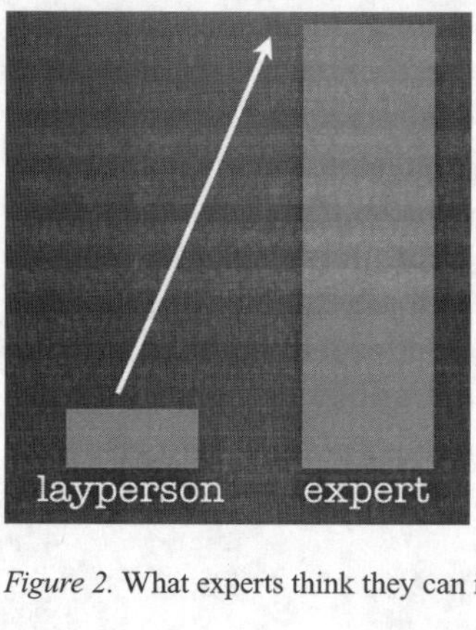

Figure 2. What experts think they can impart in one talk

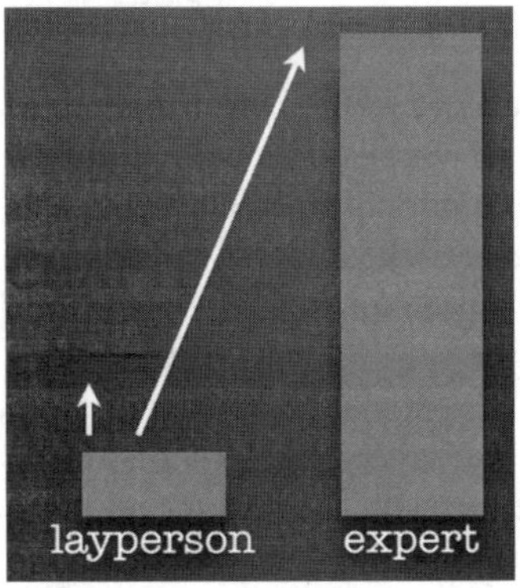

Figure 3. The reality of the situation. Aim to inform in incremental steps!

setting. I use these figures to help researchers visualize what I am saying, and continue to emphasize this as they write and revise.

2.2. The Hallway Talk

After spending several hours revising the Elevator Talk, it is important to "keep going." In most cases, it does not take long for a researcher to be able to create a fantastic Elevator Talk. However, the second one asks for more information, researchers often return to "dissertation speak" – and one can quickly watch the recipient's eyes glaze over!

For the Hallway Talk, we assume that a researcher has given a great story in the elevator. The doors open to a hallway, and the person who originally asked, "So, what do you do?" now says, "That was interesting – tell me more!"

In many aspects, writing the Hallway Talk is much easier. We have found that the Elevator Talk lends itself to being a good outline for a Hallway Talk, and that researchers can easily take each sentence from the former and expand to a paragraph each for the latter. We also stress focusing on methodology, and explaining work in the active voice as much as possible. For scientists and engineers who are grounded in writing passively, this also proves to be difficult.

2.3. Converting Writing to Speaking

After completing several iterations of an Elevator and Hallway Talk, it is easier to convert the text to a presentation. We have found that Elevator Talks convert well to a 30–60 second presentation or introduction, while the Hallway Talk can range from three to ten minutes.

Similar to what we do in the Elevator Talk training, we will go through several rounds of peer review to deal with content (and then presentation style) issues. However, the focus is mostly on refining the content so that it is understandable first and foremost.

3. OUTREACH: THE NEXT FRONTIER

Writing the complete Elevator + Hallway Talk takes time, of which researchers have precious little. While peer review within a workshop does assist in speeding up the process, it is very helpful to receive feedback from a target lay audience. As mentioned earlier, the Stanford program focused on 17–18 year old students as the audience to write to. We developed a curriculum that gave these high school and college students a chance to critically review the Elevator + Hallway Talk, with much success. Researchers found this feedback more valuable than what they received from their peers, and high school/college students learned critical reviewing skills. They also took the task seriously, as most were excited to realize that they might possibly make a difference in what a PhD-level person would say!

Another form of outreach was to have researchers talk about their work in public fora at a university. I will never forget a pilot scheme we put together, where researchers from medicine, history, and engineering came together to talk about their work. One journalist approached me afterwards and said that this was the first time that she understood enough to ask intelligent questions!

These efforts were merely a start at outreach activities, and it excites me to be able to collaborate with a group like EUSCEA, who has built a large knowledge base of such activities. My hope is that we can continue to use this training as a bridge for more scientific researchers to connect with the general public.

CHAPTER 8

"SCIENCE MEETS PARLIAMENT"

Australian researchers visit Parliament

TOSS GASCOIGNE

Council for the Humanities, Arts and Social Sciences (CHASS[1]), PO Box 8157, ANU, ACT, Australia 2601, Tel + 61-2-62491995, Email director@chass.org.au

Abstract: Australia has a problem when it comes to the relationship between science and members of Parliament. Few Parliamentarians understand the possibilities of science. They do not understand the limitations of science, or the long time scales it can take to develop an idea into something that will benefit the community. Nor do our scientists understand the work of members of Parliament. They do not have a clear idea of the political processes. They do not appreciate the pressures or the time scales Parliamentarians work to. Both sides, the scientists and the Parliamentarians recognise the importance of each other. But there is no natural dialogue between the two sides, because they come from different worlds. This paper describes three ways to build a new dialogue between scientists and MPs

Keywords: Parliament, political, Researchers, Policy, Meeting

Australia is a country of 20 million people and 226 members of Parliament; and the problem lies in the fact that only 10 MPs have formal qualifications in science.

The lack of scientific expertise can become a problem when Parliament discusses some of the big issues like water, energy, greenhouse, genetic engineering, waste disposal, or the environment. All of these issues are strongly bound up in science. Solutions and the possibility of new industries or new jobs will have a basis in science and technology.

Parliaments all over the world have to deal with these issues whether the individual MPs have any understanding of the science, or the language science uses, or the scientific method. Their lack of expertise forces Parliaments to rely on the views

[1] Further details about CHASS are available at: *www.chass.org.au*

M. Claessens (ed.), Communicating European Research 2005, 43–49.

of the bureaucracies advising them, or a few chosen outside experts. It makes Parliamentarians vulnerable to pressure from interest groups, or to plausible ideas that may have little scientific validity.

This is one side of the problem. Just as Parliamentarians understand little about science, our scientists have little appreciation of the work of a Member of Parliament. They do not have a clear idea of the political processes. They do not appreciate the pressures on an MP, from many different sources. They do not appreciate the time scales, the need for information NOW, because new legislation is to be voted on today. This will happen whether the scientists have had time to do their experiments, accumulate their results, and test through publication in the international journals.

The gap between Parliamentarians and science applies to all areas of research. There is an uneasy relationship generally between the world of research, and the world of Parliament. Parliamentarians are often irritated by what they see as the protected and privileged world of the universities and other research organisations, and the inability of this sector to provide the advice they need when they need it.

Australia has tried to bridge this gap by opening up new channels of dialogue between Parliamentarians and researchers. Two of the processes described below are currently operating, and the third is an idea we hope to introduce. The three processes are:

"Science meets Parliament"
Policy dinners for MPs
Parliamentary Science Fellows

1. 'SCIENCE MEETS PARLIAMENT'

There have now been eight annual events which bring researchers into direct contact with Parliamentarians. Six of these events have been for scientists, and they are called 'Science meets Parliament' (SmP). The other two have been for people in research and education in the humanities, arts and social sciences; and called "Humanities, Arts and Social Sciences on the Hill", or "HASS on the Hill". (The Australian parliament is built on a prominent hill.)

These events bring researchers into our national capital Canberra for one-on-one meetings with federal politicians. These people may be nominated by their university or research organisation, or they may be self-nominated. We seek a combination of senior researchers and others in the early stages of their careers. All registrants pay a registration fee, and all are responsible for their own airfares and accommodation in Canberra.

SmP is organised by the Federation of Australian Scientific and Technological Societies (FASTS).[2] It runs over a day and a half, and about 200 scientists from across the nation will register for the event. The first day is a Briefing Day, devoted

[2] Further details about FASTS are available at: www.fasts.org

to strategy, tactics and issues. Briefing Day features a range of speakers, including senior Parliamentarians and bureaucrats, journalists and successful lobbyists from other groups. The purpose of this day is to discuss ideas and to advise participants on the best way to approach MPs. What are some of the mistakes people make in meeting MPs? How should you prepare for the meetings?

Included in the day will be a panel discussion. Three MPs are on the panel, one from each of the major parties. They are asked questions by a journalist about the different approaches of their party to science and technology issues, and the most prominent concerns of their colleagues.

Another panel will consist of people who work in the offices of MPs. These people normally take notes at any meetings their MP has. At our event, they advise registrants on how to prepare for their meetings with MPs. Should they bring anything? How long will the meeting last? What is the best way to start the conversation? What will the MP want to get from this meeting?

We also invite top level bureaucrats, to explain their role in advising the Government and then implementing the Government's policy decisions. How can scientists contribute to this advisory process?

The Briefing Day may also hear advice on Parliamentary procedures, such as the committee system. It will cover practical issues: our registrants will have time between their appointments on the day and need to know the best place for coffee, or how to visit the Parliamentary Library or the Press Gallery.

It is also a chance for participants to practise a simple explanation of their work. We ask volunteers to come to the front and explain what their work is and why it is important, in 60 seconds flat. Then the expert panel – a staff member from a politician, a journalist and a professional lobbyist – will comment on the speaker and give general advice on how to approach the meetings with MPs.

The day culminates in a reception at Parliament House hosted by the Speaker of the House, the President of the Senate, and the Cabinet Minister responsible for science. All 200 registrants will attend, as well as representatives of sponsoring groups. About 50 MPs will also attend, to meet the scientists in a social environment and to hear brief speeches from the hosts.

The second day is devoted to one-on-one meetings. We arrange individual meetings between a pair of scientists and individual MPs. Normally these meetings last about 30-40 minutes, although some have stretched out to 90 minutes. The meetings take place in the office of the MP. Normally there will be four people present: the MP, a member of their staff, and two scientists.

Meeting with scientists is an optional activity for MPs. We write to them 6 weeks before the event, inviting them to participate. Usually somewhere between one half and two-thirds (113 to 150) of all MPs agree to meetings, including some members of the Cabinet (all Cabinet members are also Parliamentarians in the Australian system). More MPs want to meet natural scientists than social scientists or humanities people – this is probably a reflection of their own background in with social sciences and humanities and their assumed familiarity with issues in these areas.

Gaining this wide access to Parliamentarians is unprecedented in Australia. This is not because researchers are so highly regarded, but probably because the idea of lobbying in this way does not seem to have occurred to other groups. This is in sharp contrast to the position in the US, where lobby groups come through Congress in regular waves. SmP was originally modelled on the American 'Congressional Visits Day'; and in Washington the scientists have to contend and compete with all sorts of other lobby groups, such as 1000 teenagers wearing blue overalls emblazoned "Future Farmers of America".

What sort of issues are discussed at our meetings? There are three broad areas the conversation can cover. The first is nominated by the MP. We provide a list of twelve possible topics – like a menu – and ask all Parliamentarians to nominate issues they would like to discuss. A typical list of natural science issues follows, with the number indicating how many MPs nominated this topic. (Note that MPs could choose more than one.):

37	Water quality and Salinity
26	Energy sources of the future
26	Education and training - school, university and industry
24	Commercialisation, innovation, industry research
23	Climate, climate change and greenhouse
19	Broadband, Telstra and the bush
19	Agriculture and agribusiness
18	Health and medical issues
18	Environment and biodiversity
17	Sustainability and the triple bottom line
16	Defence science and biosecurity
16	Science in the local electorate
15	Stem cells and tissue engineering
14	Emerging technologies: nanotechnology, photonics, bioinformatics
14	Medical and agricultural biotechnology: benefits and risks
10	Nuclear power, and the storage and disposal of radioactive waste
10	Mining and resource industries and exploration
10	Brain drain, recruiting
10	Oceans and the Great Barrier Reef
9	Feral animals, noxious weeds and quarantine

A second topic of conversation is the research interests of the scientists attending the meeting, and the third is an agenda prepared by the organising body. This will focus on a big issue, such as the level of the national investment in science and research.

The meetings between scientists and Parliamentarians are arranged and scheduled by the office of the organization running the event. The events I have run initially matched a scientist with an MP according to where they lived or worked, because we wanted to link them with a politician who had an interest in them as electors. In later versions of the event, we have tended more to match by area of interest, seeking to find experts on the issues nominated by MPs.

This whole event is normally run by three or four people. Two of them are casuals employed for one or two months specifically for this event, and the others are permanent staff. We also need a good web person to manage the registration system, and a bookkeeper to handle the financial side.

There are three sources of funding: from the participants, the sponsors and the Government. The participants pay a registration fee of about € 120, and this covers most of the direct costs: hiring venues, all refreshments, and printing. The second source of funding is from sponsors. These include the research councils, some universities, even industry. We generally raise about € 40 000 from sponsors.

The Government has also provided modest support, about € 12 000 per event. The Minister is a keen supporter of the event because he has 200 ambassadors meeting MPs, and telling them what a wonderful thing science and research is. It makes it easier for him to get his budget through Parliament! Normally the event will run at a slight profit.

It is a very challenging event to run. The small team of staff has to organise speakers and venues; schedule meetings; arrange transport and printing; and deal with sponsors. MPs are not always easy to deal with, and can be slow to respond to the invitation. The volatile nature of Parliament also presents a special challenge. It is not uncommon for our scheduled meetings to be disrupted by an emergency convening of a Parliamentary Committee or some other unexpected event, and then our meetings have to be re-scheduled. Registrants are warned in advance that Parliament is unpredictable, and they cannot expect everything to run smoothly.

The event brings together a wonderful slice of Australian science, from all institutions and regions, and the participants find the opportunity to meet colleagues very attractive.

Feedback from participants has always been positive. They regularly score the overall event at about 8.2 out of 10. We ask them to complete an evaluation form, and here is a sample of their comments:

- *Very well organised. The impact is becoming obvious.*
- *I have no doubt that past events have helped put science on the political agenda.*
- *A valuable experience as a young scientist.*

We asked for separate feedback on the meetings with politicians. There was a wider range of views on these, with many reporting interesting meetings with many opportunities for follow-up activities. Others had a less productive experience

- *An excellent introduction to the person and a good first meeting*
- *MP certainly tried to direct it - said we needed to be more aggressive with media, need a publicly recognised media advocate*
- *Very positive, interested in what we had to say. Interested in nanotechnology.*
- *My MP listened but expressed no great enthusiasm for science.*
- *Of little use to science. He kept on commandeering the conversation! Warm, pleasant meeting, but no use for us!*
- *Very supportive – very useful and pleasant. Photos were taken by the adviser for the local newspaper.*

What have the events achieved?

First, they have opened a conduit for communication between the two worlds of research and politics. They remind MPs that researchers have ideas, they have solutions to problems, and they have lots of potentially-useful information. It has helped counter the sometimes negative view MPs have about people from the world of research.

The second benefit is that we are steadily building up a group of researchers who have been to Parliament and talked to politicians. This helps break down preconceived notions about politicians, and enables researchers to appreciate the pressures and constraints under which politicians operate. They have a better understanding of how they can contribute positively to the Parliamentary process.

2. POLICY DINNERS FOR MPS

A second way we are working to influence MPs is by organising a series of dinners. 25 guests are invited, half of them MPs and the other half experts in a particular field. The MPs are chosen from all political parties. This is important – we have to behave as a non-partisan organisation.

Each dinner has two guest speakers who have a public conversation in between the courses of the dinner. It is an informal conversation rather than a formal lecture, and MPs are invited to join in, to ask questions and to express their own views.

The topics are carefully chosen. They cannot be part of the current political debate, where the parties have already made up their minds and are quite often in opposition to each other. So we choose "over the horizon" topics, topics of the future where MPs are still seeking information and ideas.

The first dinner was on the topic of "Ageing Well". One speaker was an economist, the other from a medical background. They talked about what was going to happen to Australia over the next two generations, in a time when life expectancy is increasing and the birth rates are steady or declining.

The information provided was factual, and all guests were invited to follow up matters of detail with the speakers. We do not ask the MPs to do anything specific, but simply to explore an issue with a group of experts in the field. Dinners are held in Canberra during a sitting week for Parliament, when all MPs have gathered in the national capital. They are funded by the Vice-Chancellor of Australia's leading university, whose campus is close to the offices of Parliament.

3. PARLIAMENTARY SCIENCE FELLOWS

This idea for this Programme is again borrowed from the Americans, and in the US it enables about 35 scientists to work for a year in Congress as Congressional Fellows. They either work in committee system, providing expertise and advice on a non-partisan basis to committee inquiries; or they may be attached to individual members of Congress, working on a variety of tasks from handling correspondence to providing expert advice.

The programme is funded by the professional or learned societies in the US, such as the American Chemical Society. These organisations choose the scientists and pay their salaries and expenses during the year or so they spend in Washington, and cost about € 70 000 per scientist per year.

The American Association for the Advancement of Science (AAAS) administers the programme. It trains the scientists before they begin their term, and organise monthly meetings for them to share experiences and discuss issues. Many of the US Fellows are younger post-docs, while others combine it with a sabbatical year or a final contribution to their discipline before retirement. Selection is highly competitive – the AAAS has 100 applicants for the two Fellowship places it funds.

Fellows then have to choose an office or Congressman where they can work. The choice lies between a Representative or a Senator; or a Committee attached to either House. This process may take 2 or 3 weeks, with Fellows interviewing (and being interviewed by) up to a dozen offices.

Fellows need to be highly flexible to adapt to life on the Hill. They may be asked to handle incoming correspondence, or brief Congressmen in three minutes while walking from the office to the House. They need to be able to synthesise a position in simple terms, a few words, and at short notice. Issues may range far outside their field of expertise. Positions with Congressmen can involve politically partisan activities.

Participants report that this experience can change their life forever. Many former Fellows remain in Washington at the end of their term, appointed permanently to staff of Congressmen or Committees, or finding other policy positions. About one third return to their research organisations at the end of their term. Whether they stay in Washington or return, their experiences filter through the science and technology community. That community is much more aware in its dealings with government, and the way it goes about inserting science into policy-making.

Australia has not yet adopted this programme (although Switzerland did so three years ago, where it operates under Dr Margrit Leuthold as coordinator of the Swiss Parliamentary Science Fellows Programme. She is also Secretary-General of the Swiss Academy of Medical Science.)

All three approaches detailed above deal with a crucial issue, of introducing the expertise and knowledge of the research community into the national deliberations of our respective Parliaments. This audience is too important to ignore. It is not an easy group to reach, but the approaches outlined above can be adapted to local circumstances in Europe.

These are vital issues and although they may be difficult to contemplate, we have found our efforts have been richly rewarded.

CHAPTER 9

THE SCIENCE-MEDIA INTERFACE: INTERACTIONS OF SCIENTISTS AND JOURNALISTS

HANS PETER PETERS

Research Centre Jülich, Programme Group Humans-Environment-Technology, 52425 Jülich, Germany, Tel +49-2461-613562, Email h.p.peters@fz-juelich.de

Abstract: In a "media society", the science-media interface is crucial for effective science communication. Two German mail surveys of scientists and journalists show a co-orientation, but also systematic differences between both professional groups with respect to interaction-relevant expectations. In particular, scientists and journalists disagree about who should control media content. Nevertheless, the surveys prove a high satisfaction of scientists and journalists about their contacts with each other. It is unclear whether this result is limited to Germany and the thematic fields analysed (risk issues and global climate change). More cross-national and cross-thematic research is required in order to complete our understanding of the science-media interface

Keywords: Science journalism, Scientists & journalists, Science communication, Public

1. SCIENCE IN THE MEDIA SOCIETY

Often and quite rightly so modern societies are characterised as "media societies". This term points to several features, two of which are particularly important for this chapter: the media as a major factor shaping the symbolic environment of society and the media as the routine information source for subjects beyond everyday experience. While opportunities for more direct science communication – public lectures, public events, visitors programmes of scientific institutes, museums and exhibitions – exist and certainly have their value, they cannot in the least compete with the ubiquity of science in the media. As far as the general adult public is concerned and outside the educational sector, newspapers, magazines, radio, TV and increasingly the Internet

M. Claessens (ed.), Communicating European Research 2005, 51–56.

are the prime sources of knowledge about science and its relation to society.[1] The interface between science and the media thus is crucial for the public image of science and its representation in public discourses.

Two levels of this interface may be distinguished: the institutional structures and the interactions between individuals of both systems. The institutional structure consists of the organisations active in the interface and the routines for interacting with the matching system, e.g. science organisations and their public relations strategies, and editorial science departments of newspapers, magazines, radio and TV and their routines of information gathering, selection and processing. Within this institutional framework, specific interactions between representatives of the two involved systems – science and journalism – take place.

Based on empirical evidence gathered by surveys of scientists and journalists, this chapter focuses on the individual level of the science-media interface. While this chapter does not explicitly address the interrelation of the two levels, we have to keep in mind that the individual encounters between scientists and journalists are embedded in the institutional structure of the interface. Public relations and the routines of journalistic inquiry, for example, strongly influence which scientists and journalists actually talk to each other in the preparation of articles or programmes.

2. THEORETICAL PERSPECTIVES

Several theoretical approaches can help to understand the relationship between scientists and journalists – and the assessments of their outcome by both sides. One can analyse the science-media relationship as an instance of *intercultural communication* pointing to the different professional cultures of science and journalism with different languages, background knowledge, relevance systems, definitions of roles and situations, and values (e.g. newsworthiness vs. accuracy). This approach analyses communication problems as "misunderstandings" caused by cultural differences.

Furthermore, one can conceptualise interactions of scientists and journalists as *interest conflicts* from a game theoretical perspective. In most cases there is no clear antagonism between the two "players" but also no full conformity of interests. Scientists, for example, may be interested in gaining public reputation, public funding or legitimacy, while journalists are looking for a good story to attract their audience. Interactions from the interest conflict perspective are seen as negotiations in which each side aims at maximising the own utility of the outcome of the interaction "game".

Finally, we may conceptualise the relationship of science and the media using a *systems-theoretical perspective*. According to that perspective, science and the

[1] We limit our discussion of the media to "journalism" ignoring "entertainment" and "advertising" as two other content providers (cf. Luhmann, 2000) that are potentially relevant for science communication. (Consider e.g. the film "The Day After Tomorrow" for the public image of global climate change, and the many references to R & D and scientific expertise in advertising.)

media represent two different social subsystems which operate according to their own specific logic and thus cannot – in a true sense – "communicate" with each other. The media's approach towards science is based on "observation" rather than communication. The result of that observation is a media "construct" of science shaped by the operating rules of the media (Luhmann, 2000; Kohring, 2005) but also influenced by science's self-presentation towards the media. Analyses of the semantics of science communication indicate that media constructs of scientific research are not simply popularised (i.e. simplified) versions of the scientific constructs but differ systematically from these (e.g. Fahnestock, 1986; Peters & Heinrichs, 2005).

3. INTERACTION-RELEVANT EXPECTATIONS AND EVALUATION OF CONTACTS

Two recent German mail surveys of scientific experts and journalists have looked at the interactions between both professional groups from the intercultural communication perspective and provide some insights into the interaction-relevant expectations (Peters, 1995; Peters & Heinrichs, 2004 and 2005). In both studies, scientific experts who had actually appeared in the media and journalists who were the authors of those media stories were surveyed. One of the surprising results of these surveys was the finding of a strong co-orientation of scientists and journalists who are active in the science-media interface, i.e. a far-reaching acceptance of media norms by scientists and much respect for scientific norms by the journalists. Both surveys took place in Germany and in scientific fields (risk issues and global climate change) in which science provides relevant expertise for individual and political decision-making. It is thus not clear to which extent the survey results can be generalised to other countries and other research fields.

There are some systematic disagreements though. Moderate differences in the expectations of scientists and journalists were found regarding communication norms and the preferred model of journalism.[2] Scientists tend to apply scientific communication norms also to public communication. They prefer to focus on knowledge in their specialist field and – compared to journalists – they like a serious, matter-of-fact, cautious and educational style of communication. Their journalistic interaction partners do not completely disagree but look for overview knowledge, prefer clear messages, evaluative comments and an entertaining style (see Table 1, items I1 & I2, as examples).

Regarding the implicit model of journalism, scientists favour a kind of "service model", normatively expecting journalists to help them promote scientific goals and interests. Based on their professional norms, journalists, at least verbally, insist on a

[2] This summary of results goes beyond the presented empirical evidence. The table gives only some examples from longer lists of statements included in the questionnaires. For a detailed analysis of the survey data see Peters (1995) and Peters & Heinrichs (2005).

distance to the object of reporting, on their independence and on a watchdog perspective (see items I3 & I4).

Table 1. Comparison of scientists' and journalists' beliefs and expectations

		Scientists	Journalists
I1:	Experts should warn the public, even if there is only a suspicion of danger (2003)	1.2	1.6
I2:	Experts should focus on their narrow field of competence and avoid giving information about other areas (1993)	1.0	−0.7
I3:	The media should always be critical regarding environmental experts and probe into their interests (2003)	1.4	2.3
I4:	The media should accept the authority of experts and support them in the popularisation of their research (1993)	0.9	−0.7
I5:	Experts have a say in the journalistic framing of the media product, for which they have been interviewed (2003)	1.2	−1.9
I6:	The journalist should let the scientist check the article before publishing (1993)	2.1	−1.2

Note: Average agreement to selected statements from two surveys on 7-step scales ranging from −3 to +3 (positive values indicate agreement, negative values indicate disagreement); all differences are statistically significant (t-test, $p \leq 0.05$). Number of cases: 122 scientists and 85 journalists (2003 survey), 225 scientists and 136 journalists (1993 survey).
Sources: Peters (1995), Peters & Heinrichs (2005)

A clear-cut disagreement exists in the issue of control: who should control the communication with the public and the media content? Journalists consider themselves as responsible authors and scientists as their "sources", i.e. as a means to complete their task of writing a story. According to their own norms, journalists owe sources a fair treatment (e.g. correct quotation) but nothing more. In particular, they are very critical of demands from sources that may be viewed as censorship. Scientists, however, in accordance with their implicit service model of journalism, obviously think that they know best the message to be conveyed to the public, and hence should control the communication process. In both surveys, there was no other item showing such a large gap in the answers of scientists and journalists than item I5 and I6, respectively.[3]

Communication problems might be expected because of a number of theoretical reasons, such as cultural differences between science and journalism, conflicts of interest and systems boundaries leading to systematic differences in scientific and journalistic constructs of research. There is furthermore anecdotal and systematic evidence that scientists are confused by the manner in which journalism deals with scientists and science.

[3] Although journalists verbally refuse the demand of scientists to check articles prior to publication the 1993 survey showed that this form of pre-publication feedback is a frequent practice. Journalists seem to accept this practice (or even suggest it themselves) more easily if it is framed as "offer" to help improving the accuracy of the article than if it is framed as a "demand", questioning the journalists' right to control the coverage.

The general evaluation of contacts by those scientists and journalists actually interacting, however, is surprisingly positive: In the 2003 survey, 78 % of the scientists and 92 % of the journalists rated their general experiences with the respective other side as "mainly good"; the rest checked the category "good and bad experiences are balanced". Only one scientist out of 122 (and no journalist) described his or her general experience as "mainly bad" (Peters & Heinrichs, 2005). The answers to the same question in the 1993 survey were also quite positive, although somewhat less so: 53 % of the scientists and 78 % of the experts said that their experiences have been "mainly positive", 4 % of the scientists and 1 % of the journalists expressed "mainly negative" experiences. The rest of the respondents reported mixed experiences (Peters, 1995).

CONCLUSIONS

The empirical evidence briefly presented here indicates that there are groups of scientists and journalists who are able to overcome the difficulties and to interact with each other in a way leading to mutual satisfaction. One possible reason for the relatively successful cooperation is general social competence. Dealing with subcultures, systems boundaries and conflicts of interest is quite common in a highly differentiated society. We have developed strategies to deal with the problems that arise from social differentiation. These strategies – learning to understand the other culture, negotiating compromises, cynical or resigned tolerance of different perspectives and values – are obviously also effective in the science-media interface. Another possible reason for the smooth interactions is the expert role of scientists in the fields studied (risk issues and global climate change). Many scientists in these domains are not only researchers acting in the world of science but are also advisors outside science, dealing with issues of public health and policy consultancy, for example. This "expert role" probably fits much better to journalistic ways of making sense of science than the role of the pure researcher.

So far, we have treated the satisfaction of scientists and journalists as if it were a goal in itself. This is, of course, not true. The task of journalism is to serve its audience and not to make its sources happy. A meaningful evaluation of media reporting of science hence has to be based on a thorough analysis of the information needs of the audience. The satisfaction of scientists with the coverage of science is no adequate substitute for a well-developed set of evaluation criteria. In one respect, however, the satisfaction of sources is relevant: it increases the motivation to deal with the media. Our results indicate that the actual experiences of scientists with the media hardly discourage them from interacting with journalists.

More truly comparative research is needed to overcome the national and thematic limitations of the research reported here and the similar limitations of surveys of scientists conducted in several other countries so far. Recently, a multinational team of science communication researchers has completed a mail survey of biomedical scientists in five countries – USA, Japan, Germany, Great Britain and France – about their public communication activities and interactions with the media.[4] This survey will add a cross-national dimension to our understanding of the science-media interface. But even this new survey does not span the full range of scientific fields. Equally important as a cross-national analysis would be a cross-thematic analysis of the science-media interface. It could link characteristics of scientific communities (e.g. applied vs. pure research, indifference or conflict with public values) to practices and experiences of public communication.

REFERENCES

Fahnestock J (1986) Accommodating science: The rhetorical life of scientific facts. Written Communication 3:275–296

Kohring M (2005) Wissenschaftsjournalismus: Forschungsüberblick und Theorieentwurf. Konstanz: UVK

Luhmann N (2000) The reality of the mass media. Cambridge: Polity Press

Peters HP (1995) The interaction of journalists and scientific experts: Co-operation and conflict between two professional cultures. Media, Culture & Society 17:31–48

Peters HP, Heinrichs H (2004) Expertise for the public: The science-journalism interface in German discourse on global climate change. Scientific knowledge and cultural diversity. Proceedings of the public communication of science and technology network, 8th international conference Barcelona 2004. Barcelona: Rubes Editorial S.L., pp304–307

Peters HP, Heinrichs H (2005) Öffentliche Kommunikation über Klimawandel und Sturmflutrisiken. Bedeutungskonstruktion durch Experten, Journalisten und Bürger. Jülich: Forschungszentrum Jülich

[4] This survey was conducted as part of the project "Integration of scientific expertise into media-based public discourses" (INWEDIS) by a multinational team of researchers with the author as principal investigator. It is funded by the German Federal Ministry of Education and Research within the research initiative "Knowledge for decision-making processes. Research on the relationship between science, politics, and society". The analysis of the survey data is not yet completed.

CHAPTER 10

SCIENCE NEWS ON THE NET

Making scientific information traceable

BRIAN TRENCH

Dublin City University, School of Communications, Dublin 9, Ireland,
Tel + 353-1-7005668, Email Brian.Trench@dcu.ie

Abstract: The impacts of the Internet on communication of science have been pervasive but also often paradoxical. Sources and formats of scientific information have proliferated, making it ever more difficult for Internet users to discern valid information. Many scientific institutions use the Internet mainly as a means to promote themselves to particular groups, thus adding to users' difficulties. The traditional quality assurance of scientific information through peer review and formal publication is no longer effective. Those communicating science on the Internet have special responsibilities to help users, by making it clear how information have been gathered and processed

Keywords: Communication spheres, web sites, Validating information, Uncertainty

1. INTRODUCTION

Internet communication has had impacts in science that are so pervasive and so extensive that they are difficult to assess. But the process that has brought to us to this point is marked by many paradoxes:

- Seamless, global communication between scientists, facilitating information exchange and collaboration across time zones and national cultures, also brings with it accelerated specialization and intensified competition.
- The web, developed for scientists to share information, is increasingly used by research and educational institutions to address competitively prospective students, prospective employees and collaborators and broader publics of policy-makers and interested citizens.

M. Claessens (ed.), Communicating European Research 2005, 57–61.

- Professional societies and research institutions use this publicly accessible medium to target niche audiences in science and industry, as if nobody else was listening or observing.
- Public journalism formats, such as daily or weekly news updates, are used on institutional web sites aimed at niche groups, thus making their mainly internal communication externally accessible.

In the playing out of these paradoxical processes, the Internet has been adopted and adapted for science communication in ways that enhance information flows between scientists but lessen scientists' control of scientific information overall. On the Internet, the original author of a given piece of information may no longer even be discernible. In this context, we are obliged, when considering the place of science news on the Internet, to look at it from the user's perspective. From this standpoint, we can assess to what extent traditional scientific information providers meet the demands of the Internet age.

2. PROLIFERATION AND PROMOTION

The many and diverse forms of Internet-based media that have developed for the dissemination of scientific information to wider publics include modified versions of services already provided via print and broadcast media, but also formats specific to the Internet, such as portals, e-zines, forums, podcasts, news feeds, SMS alerts, video and audio clips, webcasts and weblogs. Alongside this proliferation of channels and formats, we have witnessed also an increase in the number and range of publishers of science information. These include established science media organisations but also higher education and research institutions, scientific societies, science centres and museums, public education initiatives, individual scientists, advocacy groups, amateurs, and many more social actors who have become publishers, editors and authors.

In this cacophonous space, formal and informal communication between scientists overlaps with public communication of science, and thus with science communication by non-scientists. The boundaries between previously distinct spheres of communication are weakening. Discerning valid and valuable information becomes ever harder.

The difficulty is increased by the way in which many scientific organizations use the web to disseminate information about science. Increasingly, they use directly employed science writers or communicators to ensure rapid and controlled publication of results. The formats used are often those of 'news', and the impression created, therefore, is of independent journalist activity. But the primary purpose of the 'Latest News' on the home page of Physics web (www.physicsweb.org) published by the Institute of Physics, or of 'Today's News' on the European Commission's research site is much more to promote the profile, reputation and services of the organization than to provide up-to-date information on events of public interest.

In Ireland, a colleague and I surveyed over 100 sites published by scientific institutions and found they used almost exclusively a one-way model of communication, paying little attention to the expectations of Internet users that they have the means to contribute to information and argument (Trench and Delaney, 2004).

Confirming the findings of comparable surveys in Germany and Poland (Lederbogen and Trebbe, 2003; Jaskowska, 2004), our study showed that scientific institutions used the web much more to promote themselves to professional and business audiences than to share information about their activities with diverse social groups. Facilities for feedback or forum-type discussions were found on just three sites and of these, one has ceased to exist since we did our survey.

We noted that:

- While nearly all sites provided an About Us or similar statement on the organisation behind the site, just 10 per cent identified who the site was aimed at.
- While nearly all sites provided a facility to contact the publisher, just a quarter identified a particular person for contact purposes, and only one tenth responded to a message sent to them.
- Half of the sites gave contact details for scientists, but just three sites named individual authors of individual pieces.
- Three-quarters of the sites had a News section, and, of these, nearly half had published new material in the month before viewing, but less than one third of sites overall published their content with the date of posting.
- Three-quarters of the sites had Links section, the majority of the links being to institutions of the same type as the originating site.
- Less than 10 per cent of sites had any of five identified interactivity features, including subscription newsletters, alert subscriptions and discussion boards
- One quarter of the sites provided explanations of scientific terms.

Irish science web sites, reflecting the practice of scientific institutions and agencies in other countries, communicate over the web mainly with their own institutional sectors, for purposes of promotion to peers and partners.

As a consequence of the many kinds of web publishing activity, on any substantial scientific topic there may be working papers, personal home pages, research reports, university press releases, conference papers, and formally published journal articles available on the web. In discussion groups and mailing lists, there may be commentaries and correspondence on any or all of these documents. The views of sceptics, dissidents and dogmatists may be accessible alongside each other.

In this changed communication environment, scientists cannot ensure that all scientific information reaching the public has been internally validated. Intensifying competition between sectors, institutions and publishers, along with the availability of means for much more rapid dissemination of new materials, means that the traditional standard of peer review control is greatly weakened.

An Internet search for "asthma cure" takes me to the web sites of a patient group, of companies selling cures, and of a company proposing a nutritional programme. And I have not gone beyond the first page of results. Looking for material on malaria vaccine, I find information from news media, research organisations, advocacy groups, companies selling therapies, funders and others. If I search for "nanotechnology applications", I get advertisements for reagents, papers promoting policy positions on investment in nanotechnology, and public-educational material.

The competition for attention has clear implications for media professionals working in this sphere. The distinction between journalists and non- or near-journalists is increasingly hard to make but professional journalists who would wish to see themselves distinguished by adherence to ethical codes, clearly face a particular ethical responsibility.

3. PROFESSIONAL RESPONSIBILITIES

As public science on the web grows and diversifies, web users have greater need of guidance on how to assess information and informants about science, and on how to judge reliability and trustworthiness.

The case of biomedical information is especially sensitive, as it can have specific value as a diagnosis or remedy. Web users can have access to the same databases of medical-scientific materials that are the primary research resource for medical professionals. But they may also find, in the same information-gathering process, health information from drug companies, advocacy and awareness groups, complementary medicine practitioners and mystics. There are several initiatives among biomedical publishers, such as Hi-Ethics Consortium (www.hiethics.com), and in the government sector, such as Science Panel on Interactive Communication and Health (www.health.gov/scipich), to establish standards for web sites that would allow users to discern professional, and therefore credible, sources. But it remains the case that most web users seek information through search engines that make no clear discernment between information types and sources.

An essential feature of much of the medical and other science that comes into the public domain is that it is uncertain and contested, both internally and externally to science. Furthermore, the mere fact of operating in an environment of multiple information sources and source-types reduces the scope for presenting any publicly significant information as certain and unambiguous. Thus, as more and more of the publics for science journalism have access to the Internet and encounter more and more information from more and more perspectives, ambiguity and uncertainty are set to increase.

On the other hand, the Internet, particularly through the web's hyperlinking capabilities, provides valuable means for the open, public and continuing negotiation of the uncertainties that surround us. Through this medium, the contextual information can be provided that helps publics make sense of so-called news. However, we are obliged to observe that a disappointingly small number of web-based information services have developed production strategies that realise this potential.

4. PROPOSALS

Those publishing science news on the Net can more fully meet users' needs and help them to negotiate the complex information environment in which they find themselves if they provide context for all 'news' by linking to any or all of the following: the source material, the source organisation(s), the authors, previous reports on the same topic, current reports on related topics, and different points of view on the same topic.

The linked documents listed alongside news reports on the web services of, for example, BBC (where they are identified as 'Related Internet Links') or New Scientist ('Related Articles') make superficial connections to organisations mentioned in those reports. With rather greater editorial effort, and discriminating selection, such links could help add substantial value to the reports. In this way, science news publishers can provide multi-layered presentations, allowing different groups of users to read the material in different ways, and at different levels. Hyperlinking provides the means of relating pages to one another. But merely linking to related materials says nothing about the character of the connection between two documents. Science sites should offer fuller navigational assistance to users with labels and signposts that indicate what lies behind such a link and an indication as to whether it might be worth making that link or not.

Science news publishers can also use images of various kinds to enhance the explanatory content of science news on the Net over such news in other media and they can facilitate discussion, comment and inquiries for further information through the use of the web's interactive facilities.

Searching the Net for information on science raises questions as to what and whom can be trusted. Users go back to established organisations in search of trustworthy information. Science news publishers can boost their own relations of trust with users by making their information traceable to source, using hyperlinks to improve transparency, and making themselves accessible and responsive to users' queries and comments.

REFERENCES

Jaskowska M (2004) Science, society and internet in Poland. Scientific knowledge and cultural diversity. Proceedings of the public communication of science and technology network, 8th international conference Barcelona, June 2004. Barcelona: Rubes Editorial, pp 263–267.

Lederbogen U, Trebbe J (2003) Promoting science on the web: Public relations for scientific organizations – results of a content analysis, Science Communication 24(3):333–352.

Trench B, Delaney N (2004) Public education on science: How Irish scientific institutions use the web, presented to International conference on science and mathematics education for the new century, Dublin, 23–24 September 2004.

CHAPTER 11

THE CHANGING PARADIGM OF SCIENCE COMMUNICATION: CHALLENGES FOR RESEARCHERS

From a "deficit" model to a "democratic" model

MARIE-CLAUDE ROLAND

Institut National de la Recherche Agronomique, Linguistics and Research Practices, 147 rue de l'Université 75338 Paris cedex 07, Tel +33-1-42759511,Email marie-claude.roland@wanadoo.fr

Abstract: Although it is widely acknowledged that "communicating is the doing of science", paradoxically, most researchers are seen as mediocre communicators. Obstacles to communication can be found inside the scientific community, in the training young researchers receive, in the nature of the research being conducted and in some of the values and concepts the academic culture nurtures. A new learning model is needed for researchers to develop new skills; in particular they must become reflective in order to move from a linear model of "public understanding of science" to a "democratic" model where they share with non-experts societal and scientific problems

Keywords: Writing practices; Mimetic behaviour; Connectedness; Reflective practice; Learning; Plagiarism

1. INTRODUCTION

In the wake of Michael Gibbons's "The New Production of Knowledge",[1] which described a new form of production of knowledge known as Mode II and the changes affecting the knowledge produced, the ways of producing it, the context of production and the systems of evaluation, a change in the paradigm of science communication is emerging: Science can no longer exist distinct from a society to which it would deliver wisdom and knowledge; it can no longer exist as a self-contained, autonomous enclave, for fear of being crushed down by conflicting economical and

[1] Gibbons, M. & al., The New Production of Knowledge, Sage Publications, London, 1994

M. Claessens (ed.), Communicating European Research 2005, 63–67.

political interests. The relations between science and society are dynamic, both being shaped in the interaction process. Because they must take into account the context of application, researchers must be much more aware of the implications of their research. They no longer can only bring scientific or technical answers to problems. They must take into account the values and needs of end users. As stakeholders are becoming progressively involved in the process, their aspirations and projects are taken into greater consideration. Thus researchers are asked to listen to ordinary people and to discuss the implications of their work. A culture change is needed to move from a one-way, top-down communication process to dialogue. It will no doubt present a real challenge to the scientific community, forcing it to rethink its practices, the professional skills of those engaged in research and the communication of science. In this paper, we analyse and discuss some of the obstacles which need to be overcome.

2. RESEARCHERS IN THE SCIENCE COMMUNICATION PROCESS: COUNT THEM IN!

"Science exists because scientists are writers and speakers. We know this, if only intuitively, from the very moment we embark upon a career in biology, physics, or geology. As a form of knowledge, scientific understanding is inseparable from the written and the spoken word. There are no boundaries, no walls, between the doing of science and the communication of it; communicating *is* the doing of science."[2]

"Communication *IS* the doing of science" and yet in most debates science communicators seem to be a breed apart and scientists are not among them. Moreover, scientists are generally viewed as poor communicators who need help and training in order to become "better writers and speakers" because as Commissioner Janez Potočnik recently said: "The best people to explain what is being achieved in a project are those doing the research".[3]

Clearly, the issue is about changing practices. Communication must no longer be considered as an add-on in a researcher's training and activity. Therefore obstacles impairing communication among researchers and between researchers and the public must be removed.

3. OBSTACLES TO COMMUNICATION ARE INHERENT IN MODE I EPISTEMOLOGY

3.1. Research is too Technical: Where have the Questions Gone?

We hear complaints that the research goes on without much creative input, and that all papers look alike – products of the work of large teams of super-technicians rather than that of the exciting hypothesis-driven research of creative scientists. In its quest for fast

[2] Montgomery, S.L., The Chicago Guide to Communicating Science, The University of Chicago Press, 2002

[3] Cordis News, November 2005

and prolific publication, the scientific community seems to have become short-sighted and boring, and has forgotten about intellectual excitement. Above all, scientists seem to be able to communicate science and research only among themselves, leading to a widening gap with the media and society.

Moreover, our research on the writing practices of researchers[4] shows that there may be production of a lot of data, but not necessarily knowledge.

Abstracts start with the Material and Method section: all recommendations to authors – for a scientific paper, for the abstract of a conference talk or of a poster – require that the abstract contain five key sections: rationale, objectives, material and method, results and conclusions/perspectives/implications. Our analysis of thousands of abstracts reveals that in most of them the first two sections (rationale and objectives) and the last one (conclusions/perspectives/implications) are missing. Authors start with a description of what they did, and go on relating what they found. Let us explore the meaning of such practices: does it mean that the problems and the questions – scientific and societal – justifying the research are already well known to the readers of the journal, reviewers and peers in that specific field so that there is no need to formulate them? Or does it mean that scientists do not feel obliged any longer to explain why they undertook their experiments, that only they need to know that "others" are not concerned? As for the last section "conclusions/perspectives/implications", the fact that it is also missing and that when questioned, researchers confess that it is "a difficult one to write" would tend to show that researchers find it difficult to distance themselves from their experiments and reflect about the interest and implications of their work. Many reviewers have noted in their reports that very often, authors do not seem to be aware of the interest and meaning of their research. This brief analysis thus brings us to a crucial question: has research become mostly technical and are researchers turning into technicians? No wonder then, considering such practices, that they have a hard time explaining the issues and implications of their work either to the media, to politicians or to the public!

Similarly, analyzing reviewers' and evaluators' reports, we found that one major reason why papers are rejected is that introductions are weak and discussions rambling – which again proves that researchers very often fail to properly justify their research and discuss the interest of their work. Moreover, when interviewed, researchers agree that these two sections of a paper are the most difficult to write, that they often take a long time to write because authors do not agree easily on their content. Yet these sections are key to the development of mutual understanding and provide an opportunity to show the interest of the research. Questioned about their practices and writing strategy, researchers and PhD students will say that they usually start by writing the Material and Method section and the Results section, and then, because it takes so long, they think about the introduction and the discussion. Finally, our research has also revealed that most of the time, introductions are a patchwork or catalogue of references, the meaning of which is often restricted

[4] Roland, M.C., Analysis of researchers' writing practices, PhD thesis, Grenoble University, 1995

to the hyper-specialist. A recent paper[5] lists questions often overlooked by authors and that need to be addressed, such as Why is the research done? what are its benefits? how reliable are the results? etc.

The third major observation refers to a loss of links in scientific writing: first the use of verbs like "implies, involves, plays a major (key, essential, crucial) role in, etc" is very high, especially in the introduction and discussion sections. Our interpretation is that the hypotheses of the research are underlying but that these are not made explicit to the reader. Yet the role of research is to express relations and to state clearly the nature and specific quality of the relation being explored. Difficulty to express relations can also be seen in the choice of prepositions and all kinds of connecting elements in the scientific discourse. Thus most of the intellectual work and contribution – elaboration of a question, formulation of hypotheses, and expression of relations – remains tacit, invisible.

3.2. Do Scientists Really Communicate among Themselves?

Our findings thus suggest that communication tends to be restricted to the specialists of the discipline, or to the readers of a limited number of journals likely to belong to the same scientific paradigm, i.e to know the issues and questions underlying the work presented. This means that beyond this small sphere of people sharing an understanding of the research, problems of communication quickly arise even within the same discipline, and of course between "neighbouring" fields of research. The challenge is therefore to force researchers to accept to share their questions with a larger number of peers and further, with non-specialists, the media or the public.[6,7] The pressure to publish, harsh competition within the scientific community, and practices inherited from the masters are not likely to incite researchers to make the tacit explicit.

3.3. Training: "A System of Error in Perpetuity"

"... In short many good scientists are poor writers.... Some scientists became good writers nonetheless. Many, however, learned only how to repeat all that is incorrect with respect to the prose and style of authors before them, thus establishing a system of error in perpetuity."[8]

Another specialist in scientific writing[9] adds: "Just as children acquire language unthinkingly with their mothers' milk, so scientists tend to absorb the mechanics of writing research papers unconsciously along with their supervisors' sour grapes." This system of reproduction is now widely blamed for much of the inadequate training

---

[5] European Federation of Biotechnology, Who should communicate with the public and how?, Publisher?, city?, 2004

[6] Stengers, I., Les chercheurs font l'autruche, Rebonds, Libération, 3 septembre 2004

[7] Latour, B., Défendre l'autonomie des chercheurs, La Recherche, 315, décembre 1998

[8] Day, R., How to write and publish a scientific paper, Cambridge University Press, 1991, Preface viii

[9] Meadows, A.J., The scientific paper as an archaeological artefact, Journal of Information Science 11, 1985, pp.27–30

young people receive in the lab. Knowledge transmitted is mostly tacit, implicit and apprentices tend to reproduce their masters' behaviour and habits unquestioningly. Skills are mostly acquired at the bench and are therefore overwhelmingly technical. Moreover, concepts of "neutrality", "objectivity" are hammered into apprentices' heads, which leads to disaster when it comes to writing research: use of impersonal, passive forms creates ambiguity and confusion and is responsible for a lack of quality in the research done. Many scientists complain that the scientific style has become neutral, bureaucratic and conceals the personality and character of the authors.[10]

CONCLUSION: A NEW LEARNING MODEL IS NEEDED

"[Since Max Weber] the scientific community has mostly adhered to the principle that good science must be value-free. In the meantime, more than forty years of scholarship in the history and sociology of science has shown the limits of that approach and revealed the need of contextualizing our findings and methodologies . . . Very little has been done on rethinking the training of scientists. What is needed is a replacing of the value of value-free education by a new value of educating the caring scientist."

Yehuda Elkana, President of the Central European University in Budapest

In order to overcome the obstacles which impair communication among scientists and between scientists and society, some drastic changes need to be made in laboratory practices: the "Publish or perish" system is certainly to blame as one of the main causes of the plagiarism which plagues scientific writing. The mimetic behaviour which prevails in the training of young researchers but more generally within the scientific community must be stopped and scientists must be reinstalled in their role of enunciators, allowing them to take ownership of their discourse. Some of the basic concepts and values upheld by researchers and the academic culture – neutrality, objectivity, responsibility – must be revisited to adapt to Mode II, and above all a culture of connectedness must be developed: connectedness is known to help people get smarter faster about complex issues than do competition or dumping data into people's heads! As Etienne Wenger[11] advocates, "Researchers can no longer afford to merely publish their results in hope to influence practice. (. . .) A mutual learning among practitioners and researchers [must be built]."

[10] Lawrence, P.A., & Locke, M., La Recherche, September 1997

[11] Wenger, E., Learning for a small planet project, http://www.ewenger.com/research/index.htm

CHAPTER 12

TRAINING SCIENTISTS IN COMMUNICATION SKILLS

Successes & challenges of training workshops

MÓNICA BETTENCOURT-DIAS

Department of Genetics, University of Cambridge, Downing Site,CB2 3EH, Cambridge, United Kingdom, Associação Ciência para o Desenvolvimento, Portugal, Associação Viver a Ciência, Portugal, Tel + 44-1223-766701, Email mbcd2@cam.ac.uk

Abstract: As western societies are becoming increasingly dependent on scientific and technological developments, scientists must be able to make this progress meaningful to different communities. However, science communication is still not part of the formal education of researchers. It is a key challenge to identify effective ways of training scientists in communication skills. This article gives an overview of the issues raised by putting into practice the training of scientists to communicate with a variety of audiences, ranging from the public, to the media and policy-makers, and discusses the challenges faced by Europe in the important task of promoting two-way communication between its scientists and its citizens

Keywords: Training, Communication, Scientists

1. THE IMPORTANCE OF TRAINING SCIENTISTS IN COMMUNICATION SKILLS

Western societies are increasingly dependent on scientific and technological (S&T) developments which permeate everyone's lives and day-to-day decisions: should we eat genetically modified foods? Should we use the mobile phone? Should therapeutic cloning be allowed? Pressure groups are realizing the impact of those developments on the future of humanity, and increasingly demand to have a role in defining the direction of publicly funded science and in fixing legal and ethical boundaries for the application of new technologies. Scientists should make S&T progress meaningful to different communities, thus empowering the economy and democracy of knowledge-driven societies and influencing the path of S&T.

M. Claessens (ed.), Communicating European Research 2005, 69–75.

However, science communication is still not part of the formal education of researchers. The report from the Expert group to the European Commission on Benchmarking the Promotion of RTD culture and Public Understanding of Science[1] recommended that scientists should be given training in communication skills, taking into account the need for public dialogue, debate and inclusion in decision making[1]. This has been recognized by many science funding programmes[1–3]. Countries such as the UK and Australia have training schemes and scientists who receive training report the experience to be useful[1, 3, 4]. Examples of other strategies that have been used by research councils and governments to motivate and help scientists to communicate with lay audiences are: brochures with guidelines to help in the organisation and evaluation of events[5–7]; guidelines to help in the contacts with the media[6]; media fellowships where scientists can spend some time as journalists in mass media organisations (see for example *www.aaas.org*, *www.the-ba.net*); long courses, such as Diplomas and Masters in Science Communication, amongst other initiatives.

Another approach is the use of communication training workshops. Science communication workshops are very attractive as they are generally short; they can be very hands-on, can cover a variety of different topics and skills, and they do not need to be very costly. Because of the above-mentioned properties, a workshop makes it possible to have some of the best trainers (from different parts of the world). For the same reasons, it allows the participation of strongly motivated people, from different parts of the country, from different age groups and at all stages of the scientific career, some of them in highly qualified positions. At the conference workshop on Training Scientists in Communication Skills (module I and II; WTSCS), European and Australian experts discussed and put in practice the training of scientists to communicate with a variety of audiences, from the public, to the media and policy-makers.

2. WORKSHOPS TO TRAIN SCIENTISTS IN COMMUNICATION SKILLS

Several workshops exist in different parts of the world to train scientists to communicate with the media, which is the case of the Australian workshops discussed by Jenni Metcalfe[1] at the WTSCS[4, 8]. There is a precedent from workshops to train other professionals (from business, politicians, etc) to communicate with the media and similar rules seem to apply: practice and having contact with journalists. Those workshops last an average of 1–2 days and the basis of most of them is to coach skills through practice. It is common in those workshops for scientists to prepare a press release and an interview (press, radio or TV). An important outcome is to be able to phrase the importance of their work in a short time. Another dimension of those workshops is to change the attitude towards journalists. Scientists are generally fearful or

---

[1] Director E-connect Science Communication Consultants (AU), PCST ; Jenni Metcalfe and Toss Gascoigne have developed workshops for scientists in media skills, and communication planning, which they have run in several countries for 12 years.

suspicious of dealing with the media. A number of problems in the interaction between journalists and scientists come from cultural differences between the two professions; research has shown that contact with journalists and understanding a little bit of the "other culture" may help to reduce the anxiety many scientists feel towards the media and also to become more tolerant[4, 8–10]. Thus, in many of those workshops, scientists interact with journalists and hear about the day-to-day life of journalists, their deadlines and what makes a story. Understanding how the media works also allows scientists to better assist journalists in their work and consequently to trust the journalistic product.

Few workshops exist that are dedicated to training scientists to communicate with the public. An important component of many of these is to discuss the historical context of science communication: to show strategies demonstrated to be less effective, such as the "deficit", "one-way" model of communication[11]; and to discuss new models of communication with less emphasis on contents and more on dialogue and on the diversity of audiences[12].

In some countries, training is available in speciality topics such as how to communicate risk or how to communicate with policy makers or funding entities. Increasingly risk communication and crisis management are part of scientific research. Training in those areas was first developed in the entrepreneurial and government world and has been applied to science and scientific institutions (see for example [13]). The training for communication with specialised audiences is something that has been developed recently. At the WTSCS Toss Gascoigne[2] spoke about the FASTS annual event "Science Meets Parliament"[14, 15] where more than 200 scientists and MPs meet at the Federal Parliament in Canberra. This is a two day event, in which the first day is used for training/briefing the scientists on tactics/techniques to employ when they meet the federal politicians.

Exceptions to the rule of one workshop-one audience are, for example, the European Science Communication Workshop, a workshop organised by ENSCOT (European Network of Science Communication Teachers), with the aim of equipping scientists to communicate effectively with different audiences, discussed at the WTSCS by Steve Miller[3][16–18], or the workshops "Comunicar-Ciencia" organised in Portugal[10].

3. CHALLENGES OF TRAINING SCIENTISTS IN COMMUNICATION SKILLS

Several challenges of training scientists in communication skills were raised at the WTSCS. Clearly one of them is to define what is meant by dialogue and how to train for it. In the past, communication models were followed where knowledge and

---

[2] Director of Council for the Humanities, Arts and Social Sciences (AU), PCST

[3] Head of Science and Technology Studies, UCL (UK) and Director of the European Science Communication Workshops

expertise were located mostly with scientists; this alienated the public from science and created an artificial perception of science as the producer of certainties. This view has been challenged by scientific controversies such as BSE, GMOs, MMR and global warming [11,19, 20]. It is therefore important that scientists discuss their work as it is done, so that the public has contact with "science-in-the-making", getting used to controversial facts, theories and issues. Additionally, a more interactive approach to communication has been defended recently, where social context, personal interest and lay knowledge are taken as major factors in how science is assimilated and used by the public[11, 21–23]. Dialogue should therefore involve listening to the public and understanding the nature of different audiences so that the interaction can be tailored accordingly. How does one train for this? Experience shows that discussing the historical contexts and giving examples is not enough[10]. In the WTSCS both Steve Miller and Ben Johnson[4] pinpointed the training for dialogue, and thus for listening, as one of the great challenges for the future and attributed great importance to role-playing [16, 24]. Role-playing is widely used in the Meet the Scientist Programme[24] where scientists play the audience in a scientist-public encounter. Other challenges related to dialogue are training to answer questions from the public; this may involve other developing areas such as training for risk/crisis management.

Another challenge is to give the motivation and tools to make scientists more pro-active towards organising science communication activities, for example, involving the use of the web, media or public events. The web is considered a privileged form of communication; it is easy and practical for people to popularise science via websites. Workshops may focus on how to select and organise information to put on a website. It is also important to learn how to organise an event for the public. Participants in workshops can be made aware of all the issues involved in the organisation of such events, namely identifying targeted audiences, writing projects, budgeting, fund raising, and evaluation. Finally, more adventurous researchers may want to organise activities through the media (e.g. to make a small radio program, write a news article for a newspaper or magazine). It is therefore important to give researchers the information and tools needed to be able to organise those activities and to do it in a way that promotes dialogue and that engages the public with science. However, it would be impossible for workshops to tackle all the different issues discussed here. It is therefore important to establish flexible workshop modules that can be independently attended at different times.

Another challenge is to establish good workshop evaluation practices, further on from just evaluating the process and the experience of individuals. It is very important to evaluate the long term impact these workshops have on participating scientists and on their relation with society. Have participating scientists become more proficient in communication skills? Have they become more pro-active in

---

[4] Principal Researcher and Project Developer for Graphic Science, a portfolio of innovative projects designed to take science to targeted audiences (UK)

participating/organising communication activities and more involved in dialogue with the public? Only long term follow-up evaluation schemes will make it possible to measure such impact.

4. THE ROLE OF EUROPE

In the Science and Society Action Plan, the European Commission claims that researchers, research organisations and industry have a particular responsibility vis-à-vis society in terms of providing scientific and technological information to Europe's citizens[25]. The training of scientists in communication skills is an important part of that equation. It is therefore important to take advantage of different expertises and experiences existent in this matter in different countries in Europe. Such is the role of ESConet, a European funded network of people involved in science communication, coordinated by Steve Miller [18], which aims to create a workshop to equip scientists to communicate effectively with different audiences in a variety of scenarios, taking into account the cultural differences throughout Europe. The ESConet is taking in hand several challenges addressed in the previous section, such as creating new forms of training for listening, for dialogue, for crisis management, for communicating with different kinds of public and new tools for evaluation of its workshops.

In addition to ESConet, Europe should continue to promote the sharing of expertise in this topic, through meetings, as was the case at CER 2005, and through websites. Those should allow the dissemination of experiences and expertise throughout Europe, and the influx of other expertises from outside Europe. For example, Europe can learn with experts from training workshops pioneered in other parts of the world, such as Science Meets the Parliament in Australia[15] or the Congressional Visits Day in the United States of America[26]. Websites can provide important information for scientists wanting to be more involved in science communication, such as practical guidelines (for example EC[27]), media contacts, statistics and documents that help in organisation and evaluation of events. It is important that researchers know where this information can be obtained.

Lastly, a very important aspect of improving the contribution of scientists to the public discussion of S&T is to improve their participation in communication activities. In many countries, research and teaching are the only factors that play a part in the career advancement of individual scientists[1]. An outcome of the recent Eurobarometer survey of the 25 member states was that while scientists working at publicly funded labs are seen as the most reliable source to explain the impact of S&T on society, the majority of the interviewees thought scientists put too little effort into informing the public about their work[28]. There are at least two complementary ways of promoting the participation of scientists. The first one is to optimize this participation. Europe should promote the career of science communication professionals. In a research institution these professionals should organise most science communication activities, releasing scientists from those organisational duties. This would allow scientists to dedicate most of their time to research and still participate in those science communication activities. Additionally, scientists that participate in S&T communication activities should be rewarded for this. This is rarely done at the moment. This could be done by giving importance to the presence of those activities in their CV when judging their suitability to be awarded a research job and/or research grants. Europe should lead by example in those issues.

ACKNOWLEDGEMENTS

I would like to thank José Pereira-Leal, Ana Maria Dias-Bettencourt, Sofia Jorge Araújo and Ana Paula Coutinho for reviewing this manuscript. I also thank Steve Miller, Jenni Metcalfe, Toss Gascoigne and Ben Johnson for their participation in this workshop at the CER 2005.

REFERENCES

1. Miller SCP, Koulaidis V, De Semir V, Staveloz W, Vargas R (2002) Report from the expert group to the European Commission; Benchmarking the promotion of RTD culture and public understanding of Science
2. UNESCO (1999) Harnessing science to Society-Analytical report to governments and international partners on the follow-up to the World Conference on Science
3. Pringle S (1997) Sharing science. In: Thomas RLaJ (ed) Science today: Problem or crisis? New York: Routledge
4. Metcalfe JGT (1999) Media skills workshops: Breaking down the barriers between scientist and journalists. In: Presentation to international conference on the public communication of science and technology: Berlin
5. COPUS (1996) So did it work? Evaluating public understanding of science events. The Royal Society on behalf of COPUS, pp1–15
6. OST (1996) Going public. An introduction to communicating science, engineering and technology
7. Research-Councils-UK O (2002) Dialogue with the public: Practical guidelines
8. Gascoigne T, Metcalfe J (1999) Training scientists to understand and love the media. In: World science conference, econnect: Budapest
9. Peters H (1995) The interaction of journalists and scientific experts. Media, Culture and Society 17:31–48
10. Bettencourt-Dias M, Coutinho A, Araújo S (2004) Strategies to promote science communication: Organisation and evaluation of "Comunicar Ciência", a workshop to improve the communication between Portuguese researchers, the media and the public. Comunicação e Sociedade 6:89–112
11. Miller S (2001) Public understanding of science at the crossroads. Public Understanding of Science 10:115–120
12. Select-Committee-on-S&T (2000) Third report on science and society. House of Lords: London, UK
13. Baeyaert C-P http://europa.eu.int/comm/research/conferences/2005/cer2005/presentations/14/h2_crisis_management_ workshop_cer2005.pdf.
14. Parsons W (2001) Scientists and politicians: The need to communicate. Public Understanding of Science 10: 303–314
15. Science Meets Parliament http://www.fasts.org/Fsite/SmP/AboutSmP.htm. See also article on page 43.
16. ENSCOT (2003) ENSCOT: The European network of science communication teachers. Public Understanding of Science 12(12):167–181

17. ENSCOT. http://www.enscot.eu.com/.
18. ESConet www.ESConet.eu.org.
19. Jasanoff S (1997) Civilization and madness: The great BSE scare of 1996. Public Understanding of Science 6:221–232
20. Wynne B (2001) Creating public alienation: Expert cultures of risk and ethics on GMOs. Science as Culture 10:445–481
21. Pinch T, Collins H (1998) The Golem: What you should know about science, 2nd edn. Cambridge, UK: Canto
22. Thomas J (1997) Informed ambivalence. In: Thomas RLaJ (ed) Science today: Problem or crisis? New York: Routledge
23. Millar R, Wynne B (1988) Public understanding of science: From contents to processes. Int. J. Sci. Educ. 10:388–398
24. Graphic-Science. http://www.uwe.ac.uk/fas/graphicscience/projects/training/meet_scientist.htm.
25. European-Commission (2001) Science and society action plan: Brussels, Belgium
26. Congressional-Visits-Day http://www.aaas.org/policy/cvd/.
27. European Commission Communications Guide. http://europa.eu.int/comm/research/conferences/2004/cer2004/pdf/rtd_2004_guide_success_communication.pdf.
28. European Commission (2005) Double Eurobarometer Survey: Citizens, science and technology. In: *RTD_info*, special issue November 2005.

CHAPTER 13

COMMUNICATION OF SCIENCE, COMMUNICATION IN SCIENCE

New frontiers and new models

PANELISTS: GIUSEPPE ROFFI[1], LUCIANO D'ANDREA[2], BERNIKE PASVEER[3], MILAN BUFON[4]

[1]*Consiglio nazionale delle Ricerche, EU Liaison Office. Rue de la Loi, 26. 1040 Brussels Tel +32-2-2194146, Email g.roffi@euronet.be*
[2]*CERFE, via Monte Zebio 32, 00195 Roma, Italia, Tel +39-06-3208086, Email luciano.dandrea@cerfe.org/info@cerfe.org*
[3]*Faculty of Arts and Culture, University of Maastricht, Postbus 616, 6200 MD Maastricht, The Netherlands, Tel +31-43-3883476. Email b.pasveer@tss.unimaas.nl*
[4]*University of Primorska, Titov Trg 4, 6000 Koper, Slovenia, Tel +386-5-6637732, Email milan.bufon@upr.si*

Abstract: The Panel, organised by CERFE, was focused on the changing role and features of science communication. Roffi, the moderator of the discussion, held in his introduction that science communication is to be viewed as a key component of the European research policies so that a deeper effort for its promotion is required. d'Andrea, in his speech, stressed that new models of science communication – based on assumptions different from those underpinning the Public Understanding of Science model – should be developed, also taking into account simultaneously what he termed "communication of science" and "communication in science". In this perspective, a multiple-component model of science communication, which includes communication with different categories of actors was outlined. Pasveer dealt with the main current trends in scientific knowledge production, how they affect scientific communication and which "paradoxes" they produce. Bufon's intervention was dedicated to the need for an ever increasing integration of science with the public, on the one hand, and within the various scientific communities, on the other. In this framework, he also defined a set of actions needed for promoting this integration

Keywords: Science Communication, Public Understanding of Science, Public Engagement with Science and Technology, Mediation, Social Construction of Science, Model, High Education

M. Claessens (ed.), Communicating European Research 2005, 77–80.

1. THE INCREASING ROLE OF SCIENCE COMMUNICATION

Giuseppe Roffi, introducing the panel, affirmed that the role of scientific and technological research for the social and economic development of Europe in the framework of the Lisbon Strategy is being increasingly recognized. Consequently, science communication is no longer to be viewed as an optional aspect, but as a key component in the efforts aimed at making Europe the most competitive and dynamic knowledge society in the world and as one of the key elements of public and responsible choice. This entails both an increasing investment in science communication and a deep reconsideration of tools and strategies to communicate science. In this perspective, a real integration of social sciences with the other disciplines is needed in order to support scientific and technological research and innovation, also because, in contemporary societies, the classic distinction between knowledge producers and knowledge users tends to disappear.

2. THE BLURRED BOUNDARIES BETWEEN COMMUNICATION OF SCIENCE AND COMMUNICATION IN SCIENCE

Luciano d'Andrea held that scientific communication is still mainly considered as meaning the communication of science to the public. The most influential models of science communication – that of the public understanding of science – and even the models developed after its crisis – such as the model according to which the public engages with science and technology (Public Engagement with Science and Technology) – continue to be focused on this issue. However, profound changes have occurred in the ways in which science and technology are produced and hence question this view of science communication.

As a matter of fact, many typical "structures" of science – such as the disciplinary arrangement of scientific production or the clear distinction between knowledge production and knowledge application have strongly been challenged. Moreover, new actors (fund-raisers, knowledge managers, patent experts, "eurodesigners", etc.) and new institutions (university liaison offices, science parks, science shops, etc.) perform a key function of mediation between the research world and society. In the light of these changes, separating communication to the public ("communication of science") from the communication developing in the research process ("communication in science") is very difficult to implement and may also be counterproductive. Hence the need to develop new models, larger in scope and able to take into account both sides of science communication.

Therefore a proposal has been put forward to develop a multiple-component model of science communication, which includes communication with six different categories of actors i.e. communication:

- among peers (intra-epistemic communication),
- among researchers coming from different disciplinary fields (trans-epistemic communication),

- with other actors involved in the research and innovation activities (network communication) such as research managers, technical and administrative staff, enterprises, research evaluators, local administrations, etc.,
- with the social groups (social communication),
- with political actors (political communication) and finally
- with the overall public (general communication).

3. PUBLIC PARADOXES

Bernike Pasveer held that there are currently three main trends affecting scientific and technological production. Firstly, the growing trans-disciplinary nature of science, the increasing investments of public actors in science and the ever-increasing impact of science and technology on society, which requires more attention to legitimising scientific work and results to more or less well-defined audiences/users. Secondly, state research money is increasingly being distributed on a kind of public market-place, and the better one is able to point to the direct usefulness of a study, the higher the chances one has of obtaining financing. This in itself reduces the possibilities for fundamental research or studies that contribute to a country's intellectual endowement in ways that cannot easily be translated in euros and requires universities to spend time and money on valorising their work. Finally, when dealing with science communication, it is very important that science is no longer perceived as an "ivory tower" (as in the public understanding of science model), but as a social construction entailing public involvement (as in the model according to which the public engages with science and technology).

Nevertheless, all of these trends create a series of paradoxes concerning the involvement of the public ("public paradoxes"). The first paradox is that, while we cultivate the constructivist model, we continue to practise the first model (that of the "ivory tower") which leads to superficiality and vulgarisation, and leaves the immanent normativity of knowledge production quite untouched. The second paradox is that, notwithstanding that the engagement of the public is viewed as a pivotal issue, not much effort is put into broadening 'the public' which remains made up of an "elite" of people. Finally, valorisation is assumed to be a force that works only once scientific products – knowledge, technology, etc. – leave the ivory tower of science, while the context which requires research to produce marketable, useful and relevant products appears to be a real epistemological force which affects the whole process of the acquisition of knowledge.

4. SCIENCE COMMUNICATION AND INTEGRATION

Finally, Milan Bufon argued that, against this background, there is an increasing need for integration both between science and public (providing the basis for the "communication of science") and within science (providing the basis for the "communication in science").

Integration of science entails an enhancement of the relations between higher education and research (e.g. devising strategies aimed at developing an integrated approach in both evaluation and funding systems for research, at pursuing a better co-ordination of research between universities and institutes and between high education and research work, or at stimulating and increasing mobility and efficiency of both researchers and educators) as well as an effort towards building bridges between science and society (e.g. improving the access to higher education, fostering the internationalisation of higher education and research or strengthening collaboration between higher education institutes and enterprises).

Integration in science requires the development of integrative research programmes and the promotion of research in regional and local development. This objective can be pursued through a large and co-ordinated set of actions, such as the promotion of trans-disciplinary communication, the introduction in higher education of 'industrial' or 'professional' educators, the creation of university-based enterprise incubators, career centres and a better information/communication between higher education, enterprises and local/regional communities.

CONCLUSIONS

In conclusion, the authors consider that people dealing with science communication should pay attention to the following issues:

- the communication of risks related to scientific and technological research;
- the identification of the frontiers between communication in science and communication of science;
- the use of quantitative data in scientific communication;
- the new ways of knowledge production in contemporary societies;
- the relationship among rules, bureaucracy, innovation and knowledge production;
- the impacts of science communication on market dynamics;
- the need for specific professionalism on communication in/of science;
- identification and characteristics of the public in science communication;
- innovation and science communication;
- the role of science communication in the political domain;
- science communication at a local level;
- the role of sociology and social sciences in science communication;
- the integration of social sciences with natural and physical sciences in the European Union's Seventh Framework Research Programme 2007–2013.

CHAPTER 14

ADVANCING EUROPEAN PROTOCOLS FOR SCIENCE COMMUNICATION

RODERICK HUNT

University of Exeter, School of Biosciences, The Annals of Botany Company, Innovation Centre, Rennes Drive, Exeter EX4 4RN, United Kingdom, Tel +44-1392-262360, Email R.Hunt@exeter.ac.uk

Abstract: The European Association of Science Editors (EASE) address in this chapter certain difficulties now facing editors and other professionals involved in science communication. The recent expansion of the European Union, at a time of rapidly changing publication technology, has led to an urgent need to harmonise methods and training. As the leading European organisation in this field, EASE seeks to provide access to information on best practice in science communication, primarily in the EU, but also wherever European standards in science communication are admired. The session brought together science communicators from various backgrounds with a view to identifying key areas where technical and cultural differences in methods and training might readily be harmonised into workable sets of pan-European guidelines

Keywords: Editing, Writing, Standardization, Harmonisation

In opening the session, the EASE President, Dr Elisabeth Kessler (Editor-in Chief *Ambio*, Royal Swedish Academy of Sciences), spoke about the harmonisation of methods in the publication of primary science. In research communication, we inevitably face the realities of international publishing, where a 'business as usual' model is now out of the question. OECD Science Ministers have demanded open access to, and unrestricted use of, publicly-funded research data in order to promote scientific progress and to maximize the derived value of the data.[1] There are benefits both to the scientist and to society of publishing research results in scholarly journals, and many factors influence researchers/authors in their search for such journals. The electronic delivery of scientific material has brought many new benefits. Exciting new communication

[1] http://www.oecd.org/dataoecd/42/12/35393145.pdf

M. Claessens (ed.), Communicating European Research 2005, 81–84.

opportunities now exist for researchers and publishers alike. The dynamic and mobile science publishing market needs to reflect the interests of all of its participants.

1. TRANSLATORS, NOT EDITORS

Alison Clayson (freelance writer and editor, with connections to UNESCO, Paris) spoke next on the topic of editorial harmonisation (with the assistance of colleague Dr Hervé Maisonneuve). Science and society need to build inclusive communities, thus making science understandable within many existing cultures and languages. Science needs to be made comprehensible by using accessible language which makes it useful to policy makers and citizens. Science writers bridge the gap between science and the general public, so it helps enormously if scientists themselves can speak one clear language and avoid jargon. Society at large uses many different languages and thus needs to receive "interpreted" messages, as it is usually suspicious of the scientific language itself. But communication is a two-way street, with the editor as facilitator, and needs an atmosphere of willingness to hear other views and tolerate other perspectives. Iterative negotiation and dialogue should lead to common ground, with harmonisation of reporting protocols at all levels of detail. However, such harmonisation cannot solve basic problems such as poor research methodology and weak written structure, unclear starting points, objectives and outcomes. Translators of scientific drafts often face imprecision, ambiguity, poor logic, and poor linkages, but there are limits to what can be done about these because a translator is not an editor. New technologies provide a two-edged sword: the PC is hugely valuable, but it makes us lazy (we cut and paste instead of think), it undermines linear and narrative text and, when work is transferred into PowerPoint, it gives the illusion of links. The Web, with its culture of free and uninterrupted navigation, promotes the belief that 'more is more'. In the expanded European community, the continent has become a microcosm of a world within which science can become the *lingua franca*. Professional organisations, such as EASE, form a virtual and collegial community by producing the journal *European Science Editing* and the *Science Editors' Handbook*. An EASE Web Forum discusses day-to-day issues, and workshops, training and conferences are organised. Though common standards are sought, tolerance and flexibility are also valued.

2. HARMONISING SCIENCE JOURNALS?

Professor Roderick Hunt (University of Exeter, UK) spoke next on the subject of technical harmonisation. In biological research, biodiversity is generally good and monoculture is generally bad, but is the great diversity in the macro- and microstructure of science journals an equally good thing? An example is the way in which academic publications refer to each other. In just one of the elements within a typical reference to a journal article, the statement of the author's/authors' names, it is possible to present information which differs in content, punctuation, typeface, font, and so on, until at least 48 permutations of style are achieved. Then, for each of the remaining five elements in the reference, similar numbers of permutations can also be devised. These combinations all act multiplicatively, making over one third of a billion different ways of showing the whole reference. If each of the journals currently indexed by ISI were to change to a new reference style every year, it would still take 77,000 years before all the possible styles would be exhausted.[2] Some style standards exist already. The well-known Vancouver style[3] is very sparing in its punctuation, but its abbreviation of the journal's name is unpopular and the date is not placed next to the name(s). EASE itself has proposed a standard[4] in which these faults are rectified, and a typographically-embellished version of the EASE standard,[5] the 'British plant science' standard, has been favoured by the speaker's own journal and subsequently adopted by several similar journals in its field. EASE would like to explore any other areas for harmonisation, in collaboration with interested parties.

Mrs. Jenny Gretton (Vice-President of EASE) next introduced her subject, the harmonisation of terminology. In the very earliest days of European scholarship, we did indeed manage to communicate, but this was limited by the materials available, i.e.: bone, pottery, papyrus and parchment. There was a cultural 'big bang' with the arrival of the printed word, at which time we had Latin and Greek, and also gatekeepers. Much later, the electronic methods of communication created a second 'big bang': instantaneous communication with a potentially unlimited number of readers. It helps that we already have some established norms and protocols, but there are still many issues to deal with. Languages, terminologies and nomenclatures each remain in flux and it has now become necessary for scientists to learn formally how to communicate. There is a role for editing and publishing organisations such as EASE,[6] but the increasingly international dimension of science has brought new threats to the author-editor relationship in the forms of dual-language publication, translation issues and citation indices.

3. TOWARDS STANDARDS IN SCIENCE COMMUNICATION

Professor Elisabeth Heseltine was the final speaker. A freelance editor and trainer from St Léon-sur-Vézère (France), she is particularly interested in training to promote standards, with the aim of making research reports from Europe more easily recognizable to scientists throughout all cultures. To achieve this, reports must be as concise and clear as possible. The problem is that scientists all over the world have little training in science communication. Research itself comprises planning, execution and communication, but

---

[2] Hunt R. 1994. Bibliographic references: too many variations on a theme? *European Science Editing* No.52:7–8.

[3] International Steering Committee of Medical Editors 1979. Uniform requirements for manuscripts submitted to biomedical journals. *Lancet* 1979; 1: 428–430.

[4] O'Connor M. 1991. With reference to references. *News in Physiological Science* 6:288–291.

[5] Hunt R. 1992. Editorial: standardization of references. *Annals of Botany* 70:1–2.

[6] www.ease.org.uk.

only the first two are generally taught. Well-planned and -executed research does not exist until it is also communicated. Cultural differences in science communication mean that standards are necessary for science to become understandable. EASE members offer courses in many aspects of science communication, and EASE has proposed a structural model based on published sources, editorial experience and uniform requirements for submissions to journals. The next step is to raise awareness of lack of training in science communication. Quality control of the training must be achieved, and we must supplement training with CD-ROMs, handbooks, and websites. In practical terms, European financing will be needed before a pan-European science communication and training network can be created. Members of the Brussels forum were invited to participate in this venture, either by commenting on the presentations or by responding directly to EASE.

CHAPTER 15

SCIENCE GOES LOCAL: LOCAL MEDIA MATTERS

Media relations at the Joint Research Centre's Ispra site (Italy)

ELENA CEVA[1], BERTA DUANE[2], ULLA ENGELMANN[3]

[1]*RAI, TGR, Corso Sempione 27, 20145 Milano, Italy, Tel + 39-2-31992148, Email e.ceva@rai.it*
[2]*Public Relations Unit, Directorate-General Joint Research Centre, European Commission, Ispra, Italy, Tel +39-0332-789743, Email berta.duane@ec.europa.eu*
[3]*Public Relation Unit, Directorate-General Joint Reserch Center, European Commission, 200 rue de la Loi, 1049 Brussels, Belgium, Tel +32-2-2957624, Email ulla.engelmann@ec.europa.eu*

Abstract: This chapter highlights the connections between local media and regional sites of two major research organisations: Joint Research Centre (JRC) at Ispra, Italy and Commissariat à l'Energie Atomique (CEA) at Grenoble, France. The workshop provided the opportunity for representatives of other research organisations and media to exchange experiences and define and disseminate best practices. The European Commission's Directorate-General Joint Research Centre exchanges news of its latest research results with the local media and, in turn, benefits from the dissemination of this knowledge to both local and more distant communities. Local media in turn approach the JRC for expertise on current topics. The chapter presents the views of the press officers from JRC and CEA and from the RAI local journalist

Keywords: Science communication, Local media, Perception of science, Local communities, Research organisations

1. THE PRESS OFFICER'S VIEW

The Joint Research Centre has seven scientific institutes located at five different sites.

Occupying a site of 160 ha and with some 1 700 staff members, the JRC's Ispra site in Northern Italy (Varese Province) is the third largest site of the European

M. Claessens (ed.), Communicating European Research 2005, 85–90.

Commission in the European Union (after Brussels and Luxembourg) and it hosts three scientific institutes:

- The Institute for Environment and Sustainability
- The Institute for Health and Consumer Protection and
- The Institute for the Protection and the Security of the Citizen

The site was originally built as a national nuclear research centre. It was passed over to EURATOM in the early 60s,and, later on, in the 80s it underwent a transition to non-nuclear activities.

At the corporate level, in 2005, the JRC organised 13 press briefings in several Member States and Candidate Countries and issued 20 press releases on topics ranging from the Africa Observatory for Environment and Sustainable Development to the Information Events in Turkey.

In Italy, the JRC maintains contact with some 80 media ranging from local, regional and national publications to the more specialised scientific publications. Over 70 journalists visited the Ispra site in 2005.

To elaborate further, the JRC is approached on a daily basis by Italian media:

- Local media are mainly interested in matters of immediate interest to the neighbourhood. In particular the security of the site has become a topic of great interest since there have been some rumours about potential terrorist attacks. The major local newspaper, Prealpina is read by 70% of those who say they regularly read a daily paper. This corresponds to the fact that three quarters of European citizens get their information from local or regional newspapers.
- Regional, national and specialised media approach the JRC for scientific expertise on a large variety of topics.
- Contacts with web news services (e.g. Varese News) are becoming more and more frequent and cover both site aspects and scientific news.

In 2005, two specific actions were organised in Ispra:

- An internal conference organised by the European Commission's Climate Change Unit on the Kyoto Protocol on 16 February to which local media were invited.
- Open Day on 28 May with over 1 000 visitors from the surrounding area who were given the possibility to visit the JRC labs.

Both events were well covered by local newspapers and TV, and by regional TV as well.

In September 2005, and for the first time, the JRC organised an event in the framework of the Researchers' Night and more specifically on the topic of climate change.

However, while the open days were a great success, the JRC is not leaving it at that, but finds it very important to stimulate young people's interest in science.

Over the last few years, school days were organised for classes from elementary to secondary school pupils to bring to them the challenges and excitement of the world of science.

In order to develop this idea further, the JRC is also organising classrooms for teachers and they took place in November 2005. Over 50 teachers spent half a day in small groups with scientists and were given a micro project that was then used in their lessons and passed on to colleagues.

For example, a software tool developed by JRC scientists – called V-gas – shows how personal lifestyle has an impact on climate change. This was demonstrated and explained to the teachers and they received all the necessary documentation as well as the software to set up the experiment at their school.

A students' visit organised jointly with the local school authorities in December was arranged for 530 children aged between 10 to 18.

At the end of 2005, the JRC invited local media to a press lunch during which an update on the site development (new building plans, decommissioning, waste management) was given and scientific projects with particular interest for the area were presented. This initiative should become a regular fixture (once or twice per year).

The authors wish to emphasize that media work at the JRC Ispra is slowed down because of a language problem.

Local media, in particular, have a clear preference for information in their own language.

European Commission's press releases are usually issued in English, French and German and JRC News releases only in English.

The JRC web site and most of publications are in English.

When the topic of a press release has a high potential interest in Italy, a translation in Italian is clearly necessary to obtain good local coverage.

In 2005, for example, a Commission Memorandum on a JRC study on energy consumption of electronic devices was issued in English. Since the energy issue is very high on the Italian media agenda, the JRC translated the major parts of this Memo into Italian. The impact was very good.

The chart of media coverage shows (figure 1) that despite the language handicap, the JRC and its activities are often mentioned in Italian media.

There are 225 articles and broadcasts out of a total of 1050 at the end of October 2005.

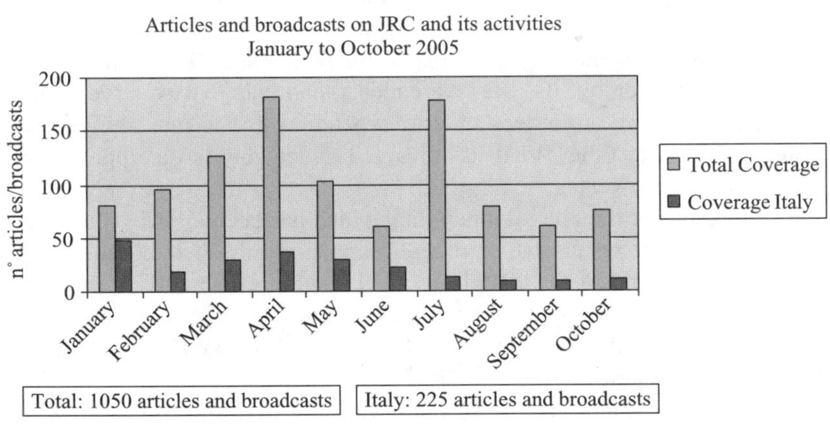

Figure 1. Joint Research Centre media coverage

The peak in January is an example of a slight media crisis caused by a linguistic misinterpretation. The JRC had made an announcement in a press release about its disaster relief work following the Asian Tsunami. The report – in English – gave a chronological summary of the event and actions taken. The part mentioning the alerts received by the Pacific station was misinterpreted by an Italian news agency. They saw this as an accusation by the JRC towards the US and the JRC Public Relations Unit was inundated with requests from the Italian media. One popular magazine even illustrated their article with an old photo of JRC's European Laboratory for Structural Assessment.

The colleagues working in this particular lab were not happy about this. Such local media echoes can have immediate impact on private lives.

One of the JRC latest press releases, however, has all the elements for success in the local and national media. The topic is the environment and climate change: more specifically, the JRC air pollution monitoring station installed on an Italian cruise ship.

It presents a strong national angle: collaboration with a famous Italian cruise line and experiments carried out in the Mediterranean. The release was issued in Italian, as well as in other languages, and the scientist responsible for the project speaks perfect Italian.

2. THE MEDIA VIEW

"Science goes local". Behind this statement, do we see just an attractive title, or a programmatic statement? In any case, it is very true that the physical presence of a scientific research centre like the European JRC within a territory does not necessarily imply a connection with local reality, beyond simple geographical location. That would be "science is local": it is here and not, casually, elsewhere. But how does science make that further step, that of moving towards the local needs and expectations of the community it is in? One way, of course is to use the press, and to let oneself be used, in no derogatory sense, by the press.

Let us examine here the effects of the journey undertaken by a specific centre, Ispra JRC, in Lombardy, Italy, in recent years. Once, all Ispra represented to the local population was the fear of a closed citadel in which – so it was murmured – foreign and unknown scientists carried out atomic experiments. It was called Euratom, and the name itself was frightening. Its gates were closed and under strict surveillance. This was the past and there is no need here to expand on it. For sure, things have greatly changed since that time. What has changed undeniably is the approach of JRC to the public and vice-versa.

The author's idea is that the open-doors policy which has been already illustrated to the public has brought results which are relevant, not only to the local public perception of science – and of the JRC in particular – but also of JRC's practical usefulness to the local community. By making itself more open, the JRC has gained in *perceived reliability* and has become useful both to the public and to local journalists, in the performance of their duties.

It is not necessary to explain here what the purpose or even the routine of local journalism implies: the sort of "magnifying glass effect" it imposes on objects, characters and events which are contained or which happen within the boundaries of its territory. It is certain that local journalists are there to amplify and to give resonance to local events. These events are not necessarily minor because they are local: they are, or tend to be, simply closer, and not just physically so, to their audience: closer to their hearts and minds, that is, or even sometimes – to their purse. It is within this framework, therefore, that we analyze the impact which a citadel of science like Ispra's JRC has on local journalism.

The JRC is, for many in the media community, a sort of huge playground: this definition in by no means disparaging, as it sometimes shows the marvels of science in a familiar, almost jocular, but in no way flippant manner. For a start, and it is of no minor effect, it surely has made science familiar, hitherto unknown in the Italian school system, which is still heavily biased towards the humanities. The study of scientific matters is often perceived as secondary, in cultural terms, to that of other subjects. All too rarely, students with a good school record are encouraged to continue in a scientific subject, and the way science is treated in school is often old fashioned – a bundle of notions and no more, as if there were, by definition, no original thought, no culture, no fun, no creativity even, in science. One way of helping Italy bridge this cultural gap is surely to show young people exactly what scientists do and who they are: not necessarily grim, boring monsters beyond the reach of human relations. On the other hand, opening doors, a policy of transparency, has a similar effect on the perception of science by the local community, young and old, who feel that researchers are working next door, but that they are not beyond the possibility of interaction, and that science does not thrive in a world totally alien to common discourse. So, it is indeed a great merit of this citadel of science that it has managed to open its gates and make itself familiar- a policy very useful to local journalists.

For a start, what is more reliable than the familiar? Unlike universities, of which there are many in Italy, the JRC research centre lets people know what they are doing and how they are doing it; its inhabitants show no snobbish restraint in speaking to non-scientists, they don't hide, as many university professors instinctively do, behind specialist language. And it is well known what the importance of language is and how hard it is at times to "translate" professional jargon, in order to make it understandable to "normal" humans!

When scientists are seen to be reliable, it makes the journalists' task easier, when they use their magnifying glass on the scientific work performed within JRC, but it also smoothes the bridging of the gap between science and local issues.

Not all scientific work produced locally is *per se* local news: what journalists need to identify is which parts of JRC production may be of immediate local interest – if the research centre is open, not only physically, this task is easier. Just one example: when Ispra decided to analyse the sources of pm 10 particle pollution, journalists knew it at once and were thus able to recount their efforts and their results, almost in real time. In a city like Milan, pollution is perhaps *the* top issue on the political agenda, and has a great impact on people's everyday lives, in terms of effects on their health, but it also weighs heavily on "hard-core" issues, such as public transport, possible limitations to the use of private cars, taxation on diesel engines or incentives for "cleaner fuel" heating systems. Journalists were able to let the public follow this research, step by step: they could see the scientists at work on a very practical, every-day issue and they could see how they formed their conclusions. Scientists were coming closer, science was going local.

Furthermore, a public service television like RAI should be able to give its local audience the means to reach an informed opinion on local issues. Pollution, climate changes, droughts, floods, land conservation, but also construction materials, or even the plastic from which children's toys are made of, are all issues with a strong, sometimes even tragic, local impact. However, when it comes to debating solutions to these problems, both political and economic factors tend to influence the policies proposed, or in any case such is the perception of the audience, and justifiably so. A politician may well be honest, but one can question his or her knowledge in scientific matters. In the same manner, solutions proposed for environmental problems, or relating to public health, imply financial efforts and economic choices and it is clear that certain parties could be suspicious that there is a camouflaged vested interest in the procedure. This is where the feeling of familiarity of the public with JRC scientists strengthens the perception of their being not only knowledgeable, but especially reliable. This is when local journalists have the chance to perform their duty of informing the public with a scientific opinion, perceived to be *super partes*: by calling JRC scientists into the arena, journalists ask them the questions they would pose to an expert, knowing that he or she does not depend economically, for instance, on local government money; that he or she is unlikely to become official advisor on that matter to one or other of the political parties. Moreover, as European research is, by its very nature, multi-national, journalists can benefit from other countries' experiences; they can try and inform the public of what has been tested elsewhere and how it has worked. And it is noteworthy that the teams working within JRC tend to be composed of scientists of various nationalities, and this works as a warrant of their independence from local economic or political powers.

Does all this mean that they are marvelous and that they always tell us the truth? Not necessarily, although journalists sincerely hope it is the case. What it means for the local press is that journalists – and with them all the local public – can ask them questions. They can rely on an authoritative extra source of information, they can continue pestering them – and this must be acknowledged, they are all very patient – and provide intelligible answers. At the same time, relying on the openness and transparency in their communications policy, journalists are able to bring scientific issues into local debate. Summing up, JRC open-doors policy has turned it into part of the community; the centre has in a sense become "our reliable next-door scientist". This, in turn means that the journalists, on behalf of the public, are encouraged to knock more and more often on their door. This cooperation seems to work. How much they enjoy it, is another matter!

CHAPTER 16

DEBATE, COMMUNICATE, EDUCATE

New media channels at your service

RUTH KIKIN-GIL

10 Lynton house, Hallfield Estate, W2 6EW London, United Kingdom, Tel +44-791-4021283, Email ruth@ruthkikin.com

Abstract: Whoever the target audience is (colleagues, big media or amateur science aficionados), whichever stage the research process is at, whatever the author's purpose is (to debate, to declare, to rant or to educate), new media channels are ready for use, and there is so much that can be done with them. Wikis, blogs, moblogs, vlogs and podcasts are all means for communicating by creating conversations between authors and their audience. They are tools for setting agendas and simultaneously informing, and being informed by the public. These tools turn the passive consumption of information to an active one, by creating communities of interest where dialogues emerge and discussions develop. These new communication channels support the curious and explorative nature of research. All that is needed is a bit of imagination

Keywords: Web 2.0, Blogs, Podcasts, Wikis, Scientific research, Research communication

The underlying question of "Science by SMS" session was "are the new technologies (such as web, mobiles, interactive TV) good or bad for disseminating science research?" The answer is complex, and is based on the ways in which researchers utilize these tools. I would like to explore the opportunities the web channels present to researchers and take a user-centred approach: How can we use technology in favour of human needs? There is a tight inter-relationship between new technologies, people's needs and social change. Each factor both affects, and is influenced by, the other two. In the case of communicating research, the technology has already changed, and I will discuss some effects later. Right now, I would like to start from the people, their needs, and their motivations.

M. Claessens (ed.), Communicating European Research 2005, 91–95.

1. ACTORS AND MOTIVATIONS

1.1. Actors

Various actors play a part in communicating research. They include researchers, research institutes, journalists, publishers, conference organisers, science aficionados and funding institutions. These actors can be roughly divided into three groups: the first group is the producers, which are the ones that carry out research, the second group is the disseminators who distribute the outcome of the research, and the third group is the consumers who are informed about the results when a research is over. That was the old paradigm (figure 1).

1.2. Motivations

The consumers are the ones whose interest is in the scientific research. What do the consumers want? Colleagues want to be informed about new research, as they need to keep up with the latest advances in their field. The financing organisations want to be convinced they gave the money to the right cause. Other motivations, which are mutual to all the actors in the consumers group, are to expand one's knowledge and understanding, to be stimulated and entertained, to discuss and to be involved.

Researchers are the producers of information, and what they want is acknowledgement of their scientific achievements. The acknowledgement enables them to gain a respected position in the scientific community (and sometimes in the public's eyes as well) and to receive further funding for their endeavours. They are also looking for places where they can exchange ideas, debate and be updated about new research. To achieve those goals, researchers must publish, attend conferences and network.

2. TECHNOLOGY

2.1. Past media

Various technologies, from print to broadcast, have played an important role in the distribution of knowledge: journals, books, newspapers, radio programmes and TV shows, to name but a few. These information channels were built on the "one to many" distribution model (figure 2).

Figure 1. The old paradigm of research communication

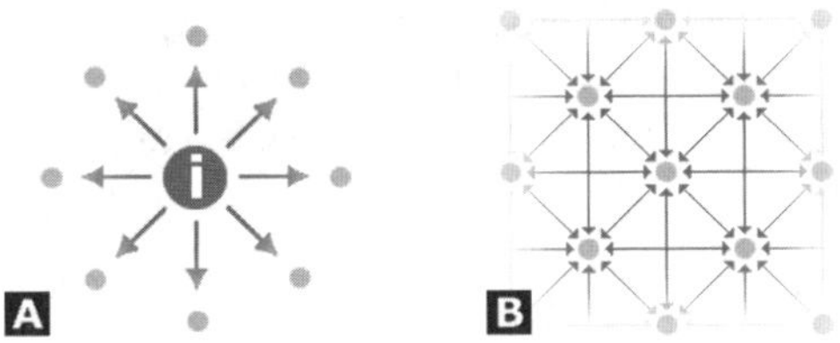

Figure 2. (A) News as a lecture, One to Many distribution model, (B) News as a conversation, Many to Many distribution model

In his book "We the media",[1] Dan Gillmor refers to these technologies as big media. In big media, news equals a lecture; one source spreads the word to everyone else. There is a clear distinction between those who create news, those who report it and those who consume the reports. In the old system, consumers had very little influence, if any, on news reporting and had no connection at all with the newsmakers.

2.2. New technologies

Before looking at the changes in the old paradigm, I would like to review briefly some of the influential and promising technologies that emerged recently. They all belong to what is commonly known as "Web 2.0".[2] The interesting aspect of Web 2.0 for us is that it transformed the World Wide Web into a platform for users to create, share and re-use content, in a decentralized way. Here are some examples.

2.2.1. Blogs

A blog is a content management system that results in an online journal with entries displayed in a reverse chronological order. Anyone can set up a blog by registering with a blog provider. Blogs vary and can be personal or professional, an individual's blog or a group's blog. Some blogs present their writers' point of view, while others are aggregators for news around a certain topic. The power of blogs lies in several features that make them a connected and reactive media: built-in feedback mechanism (comments), easy content dissemination and community awareness features (blog rolls, TrackBacks).

Blogs are easy to manage, which means anyone can write one if they are interested. Their content evolves with time and according to the author's additions and the readers' comments. The blog's format encourages conversations between the author/publisher and the readers. Blogs are an accessible publishing medium. One

[1] Dan Gillmor, We the media – Grassroots Journalism by the People, for the People, O'reilly (2004). See http://www.oreilly.com/catalog/wemedia/book/index.csp

[2] http://www.oreillynet.com/pub/a/oreilly/tim/news/2005/09/30/what-is-web-20.html

can publish by writing a blog and sharing the research process with the readers, or by letting other blog writers know about the results of a research. Once one blog has picked on the research, chances are the news will spread throughout the blogosphere, and possibly make it to big media as well.

Blogs also include built-in distribution mechanisms, called RSS[3] or ATOM[4] syndications. Users can subscribe to content through RSS or ATOM aware programmes, and receive updates about any new additions and recent changes. This means that once someone has subscribed to a site, the researcher can be certain that the subscriber will receive all the research updates.

Blogs are interconnected – a feature called TrackBacks enables anyone to see who has referred to any specific entry in their blogs. Clusters of like-minded blogs are created thanks to blog Rolls, which are lists of other blogs that the author finds interesting. This is another subtle way of raising awareness to a blog.

There are two subsets of blogs that use slightly different media, but the same principles: Moblogs[5] contain content from mobile phones (pictures and messages) and Vlogs,[6] which are video-based blogs. Just shoot your experiment on video, and share it with the world.

2.2.2. Wikis

Wikis take the collaboration features of the blog a step further and allow anyone to create and edit any page on a wiki site. The content of the wiki is the accumulated result of many minds. An example for a wiki is the wikipedia[7] – the free encyclopedia, written and edited by anyone who has something to contribute. Wikis can be used as a democratic and open platform for ideas exchange among colleagues, a place for debate and stimulation.

2.2.3. Podcasts

Podcasts are sound files that users can download from the internet and listen to with a media player or on a PC, in their preferred time and location. Many blogs, and Nature magazine[8] is just one example, offer podcasts of their content. An interesting and somewhat subversive project by Art Mobs,[9] is offering user created, unofficial audio-guides to Moma's[10] art collection, where anyone can add their comments about the works of art, download others' Podcasts and subscribe to receive the latest submissions.

[3] http://www.xml.com/pub/a/2002/12/18/dive-into-xml.html
[4] http://www.atomenabled.org/
[5] http://en.wikipedia.org/wiki/Moblog
[6] http://en.wikipedia.org/wiki/Vlog
[7] http://en.wikipedia.org/wiki/Main_Page
[8] *http://www.nature.com/nature/podcast/index.html*
[9] http://mod.blogs.com/art_mobs/
[10] http://moma.org/

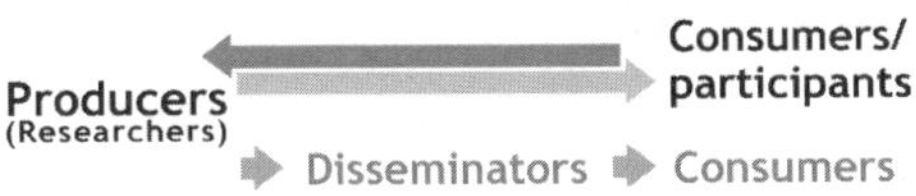

Figure 3. The new paradigm: News as a conversation

2.3. Fast media

These new technologies have evolved journalism from "news as a lecture", to "news as a conversation"[11] paradigm (figure 3). Fast media is all about accessibility, not only to content, but also to the content's creator. It supports diversity, debate and collaboration. It has become a "Many to Many" dissemination model. This shift will force the various communities of interest to adapt.

3. THE CHANGE FOR RESEARCH

The technologies previously discussed in this article made the communication between researchers and their audience more open and more direct. The options that researchers have to communicate their work have widened, and the science community can and should use them to their own benefit. Blogs and wikis are flexible media, which leave the control in the hand of the researcher to decide what they want to expose and in which format. Blogs enable researchers to start reporting about their work from the very early stages. Blogs can be a forum to consult and converse with colleagues, or rant about difficulties in the work progress. Research institutions should encourage their staff to write blogs as an unofficial communication channel with the public.

New web technologies benefit the audience as well. It empowers them by giving the opportunity to stay updated in their subjects of interest, to react and to communicate directly with researchers, to be able to follow research process and progress, and experience science (almost) first hand.

There is another facet to using these technologies: Their strengths could be their weaknesses. If anyone can write whatever they want, how can readers distinguish between the facts and the fakes? IP issues arise when publishing confidential research and the need to draw readers to one's blog may result in superficiality. All these issues are challenges to overcome. Still, the power of these new media tools should be harnessed in favor of information and participation, and it is an additional distribution channel to the existing science communication system.

[11] Dan Gillmor, op. cit.

CHAPTER 17

MEDIA SKILLS WORKSHOPS: BREAKING DOWN THE BARRIERS BETWEEN SCIENTISTS AND JOURNALISTS

JENNI METCALFE[1] AND TOSS GASCOIGNE[2]

[1]*Econnect Communication Pty Ltd, PO Box 734, South Brisbane Q 4101, Tel +61-7-38467111, Email jenni@econnect.com.au*
[2]*Council of Humanities, Arts & Social Sciences (CHASS[1]),*
PO Box 8157, ANU, ACT, Australia 2601, Tel + 61-2-62491995, Email director@chass.org.au

Abstract: There are barriers existing between the cultures and worlds of scientists and journalists. This can affect the good communication of science in the media. One method for overcoming this is through media skills training of scientists. This paper describes a successful method for training scientists developed over the past 12 years in Australia and successfully applied to other countries, including South Africa, New Zealand and The Philippines. The paper shows how these workshops appear to be breaking down the barriers between scientists and journalists and ensuring better communication of science to the general public through the media

Keywords: Media Skills, Scientists, Journalists

1. INTRODUCTION

Most people working in the field of science communication recognise the cultural barriers that exist between the scientific and media worlds. Scientists have a stereotypic image of journalists and likewise journalists have an image of what scientists are like. Both these views tend to reflect the views of the general community.

The scientific and media communities also appear to be aware of the sorts of stereotypes that exist about themselves. For example, scientists participating in

[1] Further details about CHASS are available at: *www.chass.org.au*

M. Claessens (ed.), Communicating European Research 2005, 97–102.

focus group discussions felt that the public saw them as "boring men in white coats in a world of their own, people whose actions and motives are to be regarded with suspicion or distaste" (Gascoigne and Metcalfe, 1997).

A recent survey of journalists in Australia (May 2005) where most of the survey respondents were science journalists also indicated existing barriers between scientists and journalists. The survey was sent by Econnect Communication to about 1 200 media outlets across Australia with 114 journalists responding to the web survey, most of them being science journalists (91 or almost 80%).

Most of the journalists (81%) indicated they contacted scientists consistently or sometimes. Of those who contacted scientists, 90% said they received good quality information consistently or sometimes.

A number of questions in the survey asked journalists about their view of scientists, and most journalists responded that scientists were (consistently or sometimes) passionate about their work (99%), trustworthy (97%), and helpful to the media (89%).

However, a significant number of journalists also said that scientists were (consistently or sometimes):

- were unlikely to be able to explain their work clearly without using jargon (87%)
- cared more about their peers than what the public thought (79%)
- were removed from the concerns of ordinary people (68%)
- did not understand how the media worked (62%)
- were more concerned about funding than making a difference (59%)
- were likely to hype or exaggerate their work (44%)

Scientists generally have a fear or suspicion of the media, especially if they have had little experience with them. Such inexperienced media performers "essentially distrust the media and doubt the media's potential to help their science. They are particularly fearful of misrepresentation, inaccuracy, and loss of control and see the media as exploitative and manipulative" (Gascoigne and Metcalfe, 1997).

Training in media skills can help overcome the barriers between scientists and journalists. The authors have been running two-day media skills workshops especially designed for scientists in Australia over the past 12 years. These workshops have also been run in South Africa, New Zealand and The Philippines. An initial assessment of the workshops found that "most of the media workshop graduates feel that they have better control over their media appearances, that it is helpful to their communication efforts, and that they now feel more comfortable working with the media" (Gascoigne and Metcalfe, 1997).

The authors believe an essential element to their workshops is the involvement of three to five working journalists. This paper describes how participants of 10 workshops rated the value of the workshops, and how attitudes towards journalists were changed over the course of the workshop.

It also gives some preliminary insight into how being involved in the workshops may have also helped to change some of the attitudes of journalists towards scientists.

2. THE SCIENCE COMMUNICATION PROCESS: MEDIA SKILLS WORKSHOPS

The design of these workshops has evolved over the past 12 years and each workshop is different according to the nature of the participants and journalists involved. The key features of the workshops are that they:

- are one or two-days in length and highly practical in nature (in the past five years there has been an increased demand for one-day workshops, an indication of the current demands on scientists' time)
- involve a maximum of 10 participants
- use two presenters to ensure individual assistance
- include three (one-day workshops) to five working (two-day workshops) journalists from TV, radio and print
- include interviews of all participants by each journalist

The workshops have been especially designed for scientists and technical people and are not run for any other groups in the community. A set of notes is provided to workshop participants, but it is used as a reference document, rather than as a workbook.

At the beginning of each workshop, participants are asked to list the three top things they wish to get out of the workshop from a list provided. The most popular response in every workshop is "tailoring a scientific message to suit the media, without compromising the quality of the message" (Gascoigne and Metcalfe, 2005). The least popular response is generally "understanding the pressures and constraints under which journalists work".

Each of the journalists participating in the workshop gives an informal presentation about how their particular media operates, and what they need to make a science story work for them. Demonstration interviews by journalists are given in front of the whole group, and then each participant withdraws to do individual interviews with journalists. Feedback on performance and story value is given by both the journalists and the workshop presenters.

3. EVALUATION OF MEDIA SKILLS WORKSHOPS

At the end of each workshop, participants are given an evaluation sheet to complete. These evaluation sheets are used by the presenters to constantly modify the workshop to better suit the needs of participants. However, evaluation results are always very positive, despite the initial reluctance of some participants to spend one or two days away from their research, consistently scoring above 6 out of a possible score of 7, on average.

Many (81%) of the workshop participants mention their interaction with journalists as a highlight of the workshop including statements like:

- "I liked the contact with working journalists"
- "It broke down our prejudices about journalists and exposed the areas where the talent can be at fault and can improve"

- "I liked the open discussion with journalists, and the interviews and feedback"
- "I liked the opportunity to get the inside story on how the news media think and operate"
- "The opportunity to experience interviews with different media was great – an excellent group of journalists"
- "Being able to talk to working journalists and see them as people not to be feared was the highlight".

4. PARTICIPANTS' VIEWS OF JOURNALISTS

During the first eight months in 1998, media skills workshop participants in Australia and New Zealand were asked to rate their views of journalists before and after workshops. This evaluation was based on a set of both positive and negative words provided to workshop participants, and a seven-point scale from strongly disagree (0) to strongly agree (7).

The results from 10 of these workshops were assessed according to how participants changed their views of journalists over the course of the two-day workshops. These results indicated that after doing the workshops, participants were more likely to think of journalists as helpful, thorough, concerned, reliable, accurate, trustworthy, interested and hard working.

On average, workshop participants did still tend to think of journalists as being superficial (average 4.2), with a tendency to trivialise (average 4.1) or sensationalise (average 4.7) their stories. However, participants did record positive changes to these three words and were less likely to think of journalists as being likely to sensationalise, be unprincipled or to distort, trivialise or be superficial about the stories.

5. THE JOURNALISTS' POINT OF VIEW

The media skills workshops could also be called 'scientific skills for journalists', and for many participating journalists, this is their first contact with scientists. Many of the journalists are excited about the stories presented to them during the workshops, and it is rare that at least some media coverage does not emerge from the workshops.

A questionnaire was sent to 45 journalists participating in workshops during 1998. The questionnaire was returned by 10 journalists, who were generally enthusiastic about the value of media skills training:

- "I think the workshops are extremely useful in training scientists to better deal with the media, mainly because they teach scientists to speak like 'normal' people."
- "Most of the scientists in the workshop in which I participated had never had much media contact, and they were anxious about dealing with the media. I'm sure we managed to show that really, we're quite nice people, and all we want to achieve is to be able to have a clear and concise chat about new scientific breakthroughs. Easy!"
- "Media skills workshops not only provide an important understanding to scientists of the different roles of the media but also the necessity to convey material to the public in a more understandable manner."

- "It is valuable to have people in the media meet scientists and explain how the system works."
- "They show media people as doing a job (breaks down the fear barrier), and they encourage scientists to think of the importance of their work in a way the general public can understand."

Of the 10 journalists returning the questionnaire, four had at least some contact with scientists before participating in the workshop. This included three science journalists working in the print media and on television. As such, these journalists were unlikely to have changed their views about scientists over the course of the workshop. However, some of the journalists less experienced with science stories did note some changes in their perception of scientists:

- "I was refreshingly surprised by their desire to become media savvy. All had good stories to tell and most were able to express themselves in easy to understand terminology."
- "It gave me a good opportunity to discuss various issues in more depth than usual."
- "I have found that media skills workshops have widened my outlook on reporting science and technology mainly because of my direct interaction with scientists."

All of the journalists found stories that were media worthy from the workshops they participated in, however some were unable to follow up on stories immediately due to changes in their jobs. One of the radio journalists who responded to the questionnaire also said she made some very valuable long-term contacts from the workshop. Another TV journalist said she specifically followed up a weather story on the Seven Nightly News Network and found it "an easy story to arrange, and the people involved were cooperative".

The seven non-science journalists found science difficult to report when the issue was complex or people did not explain it clearly – "the difficulty is usually breaking complex issues into something palatable and picture-friendly".

Most of the journalists (8) questioned thought that science got a reasonable run in their paper on their station. However, most thought scientists could work to improve this coverage:

- "Scientists need to have more access to workshops like yours (and not just once) and be assured of complete support from their scientific and administrative bosses."
- "Scientists need to communicate with us and let us know of developments."
- "Scientists should be more proactive in promoting/selling their stories."
- "The challenge for scientists is to find a way to make their work interesting for most people, and to feel comfortable about being more vocal about their achievements."

CONCLUSIONS

Scientists and journalists come from two different worlds. One side is characterised by a methodical and precise assessment of data from close analysis over an extended time period. The other side wants simple, direct and speedy answers uncluttered by qualifying statements. The two groups are mutually suspicious of each other.

However, it is clear that interaction with journalists over one to two days of a media skills workshop is quite powerful in changing the attitudes of scientists towards journalists. Scientists leave the workshops seeing journalists more as potential allies than as a threat to be avoided. This backs up past research by the authors which found that those scientists experienced with the media are "far less likely to be victims of the media but instead attempt to use the media to serve their personal and organisational agendas" (Gascoigne and Metcalfe, 1997).

The media skills workshops expose scientists to working journalists through informal discussions and individual interviews over an intense one or two day period. Such workshops appear to mimic the experience gained by seasoned media performers in changing the views of scientists about the media. At the very least, media training provides scientists with an appreciation of the world of journalism and the constraints and pressures under which journalists operate.

The participation by journalists in the workshops also appears to make them more aware of the particular concerns and constraints under which scientists operate.. It is highly likely that such journalists, especially the non-science general journalists, are now more aware of the scientific culture and ways to work within that culture. However, more research is needed to fully evaluate the impact of the workshops on the journalists involved.

Media skills training is an important tool for helping scientists to feel more comfortable about working with the media. It does help break down the barriers between scientists and journalists and makes each aware of the constraints and pressures under which the other operates. The breakdown of such barriers should improve both the quantity and quality of coverage of science in the future.

ACKNOWLEDGEMENTS

The authors would like to thank participants of our workshops – both scientists and journalists – for their indirect contributions to this paper.

REFERENCES

Gascoigne TH, Metcalfe JE (1997) Incentives and impediments to scientists communicating through the media. Science Communication 10(3).

Gascoigne TH, Metcalfe JE (2005*)* Media skills workshop notes for scientists and others involved in science, workshop notes, published by Econnect Communication Pty Ltd.

Metcalfe JE (2005) Personal Communication re Survey of Journalists: What do you think of Scientists?

CHAPTER 18

TRAINING FOR DIALOGUE AND DEBATE

Some challenges

STEVE MILLER

Department of Science and Technology Studies, University College London, Gower Street, London WC1E 6BT, United Kingdom, Tel +44-20-76793490, Email ucapt0s@ucl.ac.uk

Abstract: Science communication training for natural scientists and engineers has traditionally focused on basic skills, simple communication techniques, and straightforward interactions with the media. In its *Science and Society Action Plan* of November 2001, the European Commission outlines 38 actions designed to bring the development of science and technology into line with the aspirations of European citizens. Many of these involve scientists engaging in dialogue and debate with the public, either directly or through the media. These are activities for which most young scientists receive little (if any) formal training. The challenges for science communication training that the *Action Plan* implicitly involves need to be enunciated and discussed

Keywords: Science communication, Training, Dialogue and debate

1. INTRODUCTION

Since the mid-1980s, scientists and engineers have been encouraged to discuss the work that they do with the public. A variety of reasons have been advanced for this. The need to overcome public hostility to science, the need to equip citizens with the knowledge they require to live in an age of modern technology, and the need to integrate science better with the overall culture of western civilization are among some of the more common arguments advanced for improving what has been called "scientific literacy" or "public understanding of science". But there are many other arguments that have been put forward, with more or less conviction. Recognising that there is not much in the general education of researchers that equips them for such communication activities, many individuals and bodies

M. Claessens (ed.), Communicating European Research 2005, 103–107.

have developed training schemes designed to fill the gap. As a result, there are several groups who run short workshops, and there are now many fully-fledged degree programmes – at both post-graduate and undergraduate level – that have science communication as part or all of their goals. All of these attract natural scientists and engineers as participants, and there is no doubt that the need for communication skills is much better appreciated in the scientific community right across Europe than was the case a generation ago.

But just as the science communication activities of the mid-1980s and early 1990s have been characterised as emphasising the deficit in public knowledge and the superiority of scientific information – the so-called Deficit Model – so training schemes have largely been aimed at the one-way communication of information. That is no longer enough; we are in a science communication climate that the British House of Lords described as "the new mood for dialogue and debate". This is a challenging climate, both for the scientific community and for those who wish training researchers to meet those challenges.

2. TRADITIONAL SCIENCE COMMUNICATION TRAINING

Many of the people who started science communication training activities came traditionally from the media. Journalists and broadcasters, particularly those who covered the science beat, felt that they had insights from which researchers could benefit. At the same time, putting young scientists into simulated media situations – writing short news articles or press releases, being interviewed, giving a press conference – was seen as a way of making basic communication training more interesting than taking them through the nuts and bolts of the Laswell model and its successors. The emphases of these science communication workshops were on simple, clear communications, thinking about what would make the information to be communicated interesting to the "ordinary citizen" (or journalist/broadcaster standing in for this citizen), and learning a bit about the person to whom the information was to be communicated. In particular, these workshops often introduced scientists to the constraints under which journalists and broadcasters worked, and made them more sympathetic towards – or at least, understanding of – the problems and skills involved in working in the media.

In some instances, researchers were also introduced to issues of communication with policy- and decision-makers. For example, one European project, the European Network of Science Communication Teachers (ENSCOT, 2003), ran workshops that involved putting participants into "scenarios" in which they had to talk about their science to members of the European Parliament or European business leaders, as well as the more "traditional" science communication activities. All of these are worthy and important skills for researchers to acquire; much of the communication that they have, or will have, to carry out will involve them explaining matters in which they are experts to those who are not, and maybe answering a few questions to clarify what they have said or written. Within the "new mood for dialogue and debate" the skills involved in clearly explaining complex scientific and technical issues are absolutely vital.

3. THE "DIALOGUE AND DEBATE" CLIMATE

In the late 1990s, it became clear that the Deficit Model of science communication, which involved a one-way, top-down approach and characterised the general public as a passive, rather ignorant, amorphous mass, was, at best, inadequate. Crises and scandals involving mad cows, impure cooking oil and infected blood had shaken confidence in the assurances of experts, particularly those that were seen as spokespersons for governments and business interests. Throughout Europe, a much more questioning approach to science and technology, and the benefits they purported to bring with them, had developed. In 2000, the UK's House of Lords produced its report on *Science and Society*, which called for a "presumption of openness" regarding the advice that experts gave to the government of the day, and insisted that public participation in the development of science policy be carried out in good faith.

In November 2001, in the wake of the latest survey on the attitudes to science and technology, the European Commission's *Science and Society Action Plan* argued "there are indications that the immense potential of our [scientific] achievements is out of step with European citizens' current needs and aspirations, such as peace, jobs, security and sustainable development of the planet". The Commission set forward a three-fold strategy: to promote scientific and education culture in Europe; to bring science closer to citizens; and to put responsible science at the heart of policy making. Many of the programmatic actions contained in the *Plan* – such as the proposal for local and regional "science and society" forums – involve scientists engaging in dialogue with the general public.

4. RISK AS A DRIVER FOR "DIALOGUE AND DEBATE"

It is clear that there has been evolution in the way in which science communication is conceptualised – from the mid-1980s Deficit Model to today's science-in/and-society approach. Questions of risk have often been at the centre of the "scandals" that have driven this evolution – "we [the public] weren't told about the risk of eating beef/having a blood transfusion etc. So we [the public or their media "representatives"] don't trust you [the scientists] any more." At the same time as the modern "public understanding of science" movements were picking up momentum in the 1980s, sociologists were looking at developments in modern society around the perception and actuality of risk. One of the most influential was Ulrich Beck, who coined the term "risk society" (Beck, 1986). This, he said, was characterised by concerns over the production and distribution of goods and services giving way to concerns over the production and distribution of *risk* – with the question of who was producing risks for whom, with or without their knowledge and agreement.

Risk, for Beck, had implications for science communication. While experts believed that people would not worry if they had the facts, as they were known to the expert, Beck countered: "Even in their highly mathematical or technical garb, statements on risk contain statements of the type *that is how we want to live.*" And he added, rather prophetically: "[I]n their concern with risk, the natural sciences have disempowered themselves somewhat, forced themselves towards democracy." The implication of this for science communication is that, by its very nature, risk, associated with scientific and technological development and assessed by scientific and technological criteria, put science firmly into the arena of public discussion. Once there, it could be interrogated according to the norms and practices of the general public, which would not privilege scientific discourse over other forms of communication. Taking the issue of risk communication further, other workers have pointed to the way that processes of "translation" are required to enable citizens to understand – into a sequence of events, into everyday experiences, and into concrete action, with accompanying solution-finding and lame-apportioning (Weingart *et al.*, 2000).

5. SOME ISSUES AROUND "DIALOGUE AND DEBATE"

Dialogue and debate, as an approach to science communication, public engagement with science and the development of "responsible" policy making, is clearly welcome. It is a development brought about partly by academic criticism of the Deficit Model and partly as a pragmatic "it isn't working, let's try something new" response to real problems. But it is not a panacea and should not be seen as such. Many issues remain to be addressed. There is an overwhelming sense in much of the discussion around science and society relations today that by involving experts, researchers and citizens in dialogue that a consensus will be reached on how to proceed with scientific and technological development. In Europe, however, we have a tradition of democracies in which issues are contested and majorities reached, which often leave sizeable minorities unsatisfied. So one can ask: "Why should this be any different for science and technology issues? Why do we expect consensus to be reached?" Given the social, political and historical variations across Europe, consensus may be reached more easily in some countries than others – one size may well not fit all.

But even if we accept, as a working hypothesis, that consensus is both desirable and achievable, one can then ask: "Whose consensus is it? And *how* is it to be achieved?" Taking the second question first, much official writing assumes that the media (somehow transformed, perhaps) will play an important part in this. But the media have their own agenda, values and practices. In particular, they do not welcome consensus. "Chirac and Blair agree on everything" might be a surprise, but it is not really an attention-grabbing headline. It is much more likely that "Chirac/Blair slams Blair/Chirac over . . . " (fill in almost any subject you want) will make it onto the front page. The media prefer conflict to consensus, and if one newspaper sells copies by supporting one side, it is a safe bet that another newspaper will seize the opportunity to support the opposing point of view in order to attract readers.

Of course, the (usually) privately-owned, commercially driven mass media are far from being the only agents for consensus-building. A battery of mechanisms – consensus conferences, focus groups, deliberative debates, opinion polls *etc.* – is available and nowadays used more and more frequently. But there are criticisms that "dialogue and debate" has been appropriated to the old Deficit Model approach as a better way of persuading citizens that what scientists want to do, and what big business thinks is profitable, is what we *should* do. In other words, this view of dialogue involves dua-logue but mono-aud: both "sides" are allowed to talk, but only one side – that of the general public – are supposed to listen, take note and change. The scientific and business communities listen only to work out better persuasion strategies – not to take note and change in response to citizen aspirations. So is this what we want from science and society relations in the 21st Century?

6. CHALLENGES FOR SCIENCE COMMUNICATION TRAINING

The preceding discussion has considerable implications for science communication training, if we want a genuine climate of dialogue and debate that allows for citizen concerns to influence the direction of technological development, without allowing research programmes to be utterly subject to the whims and prejudices of media barons. It is clear that the basic skills training workshops currently available remain absolutely essential: understanding and communicating with the media, clear written and verbal communication and straightforward presentation skills are *sine qua non* for making yourself understood. But what is taught now will have to be much more clearly situated in understanding the development of science and society relations than has previously been the case. Researchers should know at least the outlines of the deficit-to-dialogue story; they might also like to understand that there are differences between scientific literacy, public understanding of science and *culture scientifique*.

Training workshops will still work well by making use of scenarios – writing a news article or press release, simulated media interviews, presentations to policy makers *etc.* But new courses involving the communication of risk, discussions of solutions to real-life citizen problems, and – perhaps most importantly – how to *listen* are now needed. In the ESConet Workshop project, funded by the European Union's Sixth Framework Programme (2002–2006), we are developing such modules, with a view to offering training to researchers on networks funded by the EU. Work is well underway, and the first workshops will be offered from September onwards. Watch this space!

Steve Miller is Professor of Science Communication and Planetary Science at University College London, and Director of the ESConet Workshops project.

REFERENCES

Beck U (1992) Risk society: Towards a new modernity. Sage, London

The ENSCOT Team (2003) ENSCOT: The European network of science communication teachers. Public Understanding of Science **12**:105–108

European Commission, science and society action plan (CEC Brussels, 2001)

House of Lords. Science and society. London, 2000

Weingart P, Engels A, Pansegrau P (2000) Risks of communication: Discourses on climate change in science, politics and the mass media, Public Understanding of Science **9**:261–283

CHAPTER 19

TRAINING SCIENCE COMMUNICATION IN A SWIFT MOVING SOCIETY

GEMMA REVUELTA

Science Communication Observatory, Universitat Pompeu Fabra. Passeig de Circumval lació 8, 08003 Barcelona, Spain, Tel +34-93-5422446. gemma.revuelta@upf.edu

Abstract: Inverting the order of the elements in the title this chapter will first explore the idea of "a swift-moving society". It is not just about changes and opportunities derived from the information and communication tools (which have been provocatively labelled SMS synecdoche by a speaker at the CER 2005 conference). It will also revise some features of the knowledge citizen: impatient to be informed, supplied with tools (in the best possible scenario) but in many cases with a severe lack of capability to discriminate serious and trash information. It will then briefly look at the impact these changes are having on communication in general and, more specifically, the communication of science and technology. Finally, it will comment on a few ideas for a new type of training for scientists that includes communication as a fundamental part of their skills, not only in the sense of being able to relate to the media, but also in the broad understanding of emerging opportunities of direct communication with society. Bypassing conventional media and communicating directly with the citizens is one of the future challenges for the new scientists of the knowledge society. For such a task, the Internet could undoubtedly have a prominent place, but we might also think about other tools, such as SMS, for instance

Keywords: Training, Science communication, New technologies, Information and communication technologies, SMS

1. A SWIFT MOVING SOCIETY

The swift advance of science and technology and, particularly, the diffusion of information and communication technologies have such an impact on the way we work, enjoy our leisure and relate to one another that few now doubt that we are facing a social revolution.

M. Claessens (ed.), Communicating European Research 2005, 109–113.

We call it the knowledge revolution, in the hope that it is not just technology – that is, bits – that lead to the change. We say knowledge, because we wish to believe that the real change will be led by people – that is, their neurons! In other words, we hope that, in the knowledge society, bits and neurons will replace the chimneys that were the symbol of the industrial revolution.

This is the ideal. But are we really moving towards this idyllic knowledge society? What is actually happening in the field of science and technology?

Let us briefly look at some data that is very illustrative of what is going on. The responses to a survey of the adult population of the United States regarding their main sources of information on science and technology[1] showed:

- When asked "what is your main source of information on science and technology?" the most frequent answer was television (44% of cases)
- But when the question was "what source do you most consult for information on a specific scientific subject?" Here, television is not the most commonly used medium. Nor are the other conventional media such as radio or newspapers. To find out more about a specific subject in the field of science, the citizen representing the knowledge society looks, particularly, to the Internet (44%) and, in second place to books (24%).

The switch in positions of television and the Internet in terms of passive access to information as opposed to an active search is highly significant. We will return to this point later.

However, this data also highlights that the appearance of the Internet and all the other social changes of recent decades (especially in Western countries and the Northern hemisphere) has also transformed the role of the other communication media.

2. A MASS MEDIA PICTURE

Some of the main features of the current mass media landscape derived from the knowledge society can be summarised as follows:

- Financial transactions, market changes, takeovers, mergers, agreements, etc., have all painted an apparently complex and diverse picture of the media, but the truth is that there has been a practical monopoly of communication by a handful of business groups that have divided up the business in all possible formats: daily press, magazines, radio, television, etc. This leads to a clear homogenisation of information often linked to the tendency to cut costs (e.g., the same journalist who does a report for the newspaper also records it for television and radio) and to increasing market pressures (not so much in terms of quality, but in terms of immediacy of information). All this does not exactly contribute to better communication of science in the conventional media.

[1] *Science and Engineering Indicators 2002. National Science Board. 7–34.* Arlington, VA

- Added to this is a general reduction in the consumption of the conventional media, especially the daily press. This tendency is not new, but it has been exacerbated by the Internet and, among young people, by the time taken up with electronic games.
- We therefore have the scenario of the traditional competition between media (between press and radio, between radio and television, between television and the Internet . . .).
- There are also two new players on the scene: the free press and the local television. Free press has tremendous circulation among young people, who are delighted to be given a free newspaper on the underground, which they can even have a quick read of on their way to school or work but, so far, without generating a habit. That is, young people who read the free press today will not necessarily buy newspapers as adults: they will not be willing to pay for newspapers and they read them because they have been given them. Local televisions are also a new phenomenon in some European countries (like in Spain), and probably with a very high potential but, in many cases, limited by scarce resources.

3. SCIENCE IN THE MEDIA

If we add to this picture the fact that science and technology are better covered if the journalist is a specialist and if the work is done in depth and without haste, we are clearly in hard times for science – at least for communicating it in the conventional media.

In the written press and in the newspapers in particular, the expected boom in scientific journalism with the appearance of science supplements in some of the big newspapers has followed very different paths. But it has not become generalised. There are exceptions because, fortunately, some of the big newspapers continue to put their weight behind science. The most recent example is the French newspaper, *Le Monde*, whose new format includes a section entitled "Environnement & sciences". The section is also in a preferential position, after "International" and before "Europe", "Politique", "Société", "Economie", etc.

Nor is the science content of programmes in radio and television spectacular (also with honourable exceptions)! On many occasions, this is because the radio and TV stations start out with small budgets and minimum costs and it is difficult to make this compatible with the right specialisation.

This is the situation of the conventional media. Let us return to the Internet. We talked about the active search for information and passive access. What do we mean by active search? Here is a clear case in point.

The day the media announced the withdrawal of Vioxx (a drug widely distributed for the treatment of arthritis that turned out to have considerable side effects), thousands of patients all over the planet searched for additional information. Google provided long lists of web sites on the subject. In the best case, citizens of the knowledge society did not go to Google to search for information, but went to the web sites of organisations they trusted. The problem is that in some cases there is a delay between media information and official reaction.

In Spain, for example, information about Vioxx was not posted on the site of the ministry for health until 24 hours after media talk about. And the citizens of the knowledge society are not patient. They know how to use the tools of technology, but they want immediate answers.

So in the best case, the main information came directly from the pharmaceutical laboratory. The main information was usually obtained from the pharmaceutical laboratory directly who found themselves in the delicate position of having to provide real, accurate information which could possibly have the effect of compromising the sales of their own product. The availability of the tools provided by information and communication technology, together with the impatience so common in our time, where we want everything now (perhaps obsessed by the view (:o) that nothing lasts: no permanent jobs, no marriage for life, no scientific truth that stands up for more than a couple of days . . .). This urge to get information immediately, combined with the availability of the tools provided by ICTs and with a generalized lack of critical outlook, can be misleading:

- Thus, if you *google* "evolution" you may end up on a page of the journal, *Nature*, or on the Intelligent Design Network . . . or even on the Vatican website.
- And if you search for "HIV", you can find messages as different as "HIV is after you, please play safe" (i.e., use a condom) and others that say "AIDS doesn't exist" (i.e., don't use a condom).

4. TRAINING SCIENCE COMMUNICATION

Today, there is a need for scientists to play an active part in the process of social communication of science. The media play an active role in communication but that, in the knowledge society, there are other actors. And there are also the needs . . . and limitations of thc knowledge citizen.

Scientists must communicate. And they should do it well. In fact, communication should be one of their basic functions, as much as thinking, testing, experimenting . . . Ideally, scientists should learn to communicate with society at all stages of their career.

With this idea in mind, a course in science communication was created in my university in the fifth year of biology. And I am proud to be the person in charge of it. Throughout one term, I try to get my students to understand that communication is part of their job and that they should therefore try to do it as well as their other functions. The main features of the course are the following:

- The course reviews the fundamental aspects of the relationships between society and science: public understanding and perception of science in general or of specific aspects, such as biotechnology, society's participation in scientific decisions (for example, political decisions regarding research priorities).
- A large part of the course concerns relations with the media. But the idea is not merely to learn how to write a press release, but for the students and future scientists to understand the requirements of journalists and their media, their timeframes, their limitations and the manner (singular is best) in which they work. And that they learn that the best relationships are built on mutual knowledge and respect.

- Students learn oral and written communication skills and also specific aspects of communication, such as communication of risk information (always closely linked to science) or communication in crisis situations.
- All these components are necessary and useful, but the part that I am particularly convinced will be of most use to them is where they work on the different forms of direct relations with the public.

There are many examples of this type of new, direct communication between scientists and citizens. One of them – well known in the field of medicine is the web page of a scientific journal (e.g. the Journal of the American Medical Association), with a specific profile for patients, who make up an extremely high percentage of the visits to this page. Another example could be the website www.visindavefur.hi.is whose intention is to create a science encyclopaedia, not using the subjects chosen by scientists, but the questions asked by the readers . . .

JAMA and Visindavefur websites could be considered as two types of new forms of direct communication between the scientific community and society, bypassing conventional media. Others will doubtless appear in the coming years: and one of them could be – why not? – communication via SMS . . .

CHAPTER 20

SCIENCE DAYS – CONTACT WITH SCIENCE

Science in a theme park – pulling crowds for research with public events

JOACHIM LERCH

Foerderverein Science und Technologie e.V., Mundinger Weg 33, D-79331 Teningen, Germany, Tel +49-7641-933214, Email j.lerch@t-online.de

Abstract: This report describes "Science Days", a local science festival in the southwest part of Germany. It has been taking place since 2000 and brings in annually more than 20 000 visitors. One of the main target groups are pupils. The festival is planned and then takes place in co-operation with scientists, engineers, teachers and people from other occupational groups. The report also describes the manner in which the organisers improve the quality of the festival

Keywords: Science festival, EUSCEA, Science communication, Science education, Science event

1. WHY SCIENCE DAYS?

Let us start with a short story. A little boy asked his father:"Do fathers always know everything better than their children?"

The father answered:"Yes"

The boy:"Dad, who invented the steam engine?"

The father answered, while at the same time being extremely relieved that he could answer this question with"James Watt".

The boy:"But why didn't James Watt's father invent the steam engine?"

This story shows how important it is to support children and youngsters for the new generation of academics in natural science and technology. Science Festivals have proven to be especially suitable to introduce science and technology to young people. In addition to being offered the possibility to conduct experiments and constructions themselves, they also have the opportunity to meet and talk to scientists in person.

M. Claessens (ed.), Communicating European Research 2005, 115–118.

These meetings often occur in ordinary places, for example, the supermarket, park or train station rather than at scientific institutes.

2. FACTS

"If the people don't come to science, science will have to come to the people". This was the motto of the *Foerderverein Science und Technologie*, at the time the first "Science Days" event took place in Germany's biggest and most successful recreation and family park, the "Europa-Park".

The first two days of this three-day event, mainly focused on schoolchildren. On the last day, a Saturday, the entire public was invited. In 2005, there were 28 000 visitors and the total number of visitors has now doubled since its foundation in the year 2000.

Approximately 100 institutes in the field of science, industry, education and administration participate in the Science Days event every year. Among them are also educational institutes for teachers and schools (see figure 1). In 2005, more than 2 200 employees of these partners participated. Workshops, demonstrations, radio stations and science shows are presented at 100 booths which are set up over 5 500 m^2 indoors and 12 000 m^2 outdoors.

Enterprises and foundations are the main financial sources for this event. The Baden-Wurttemberg Ministry for Culture, Youth and Sports provides the necessary personnel.

The Europa-Park is a very professional environment for this event. Without the Family Park's intensive support during the preparation phase and the actual event, a festival of this size would only be possible with considerably higher funds. The Family Park is therefore the biggest sponsor of Science Days.

3. THE THEMES

The organisers very deliberately value a great variety of themes. Although different themes of emphasis are selected each year (in 2006, for example, forensic and information technology have been chosen), any other themes concerning science and technology are acceptable.

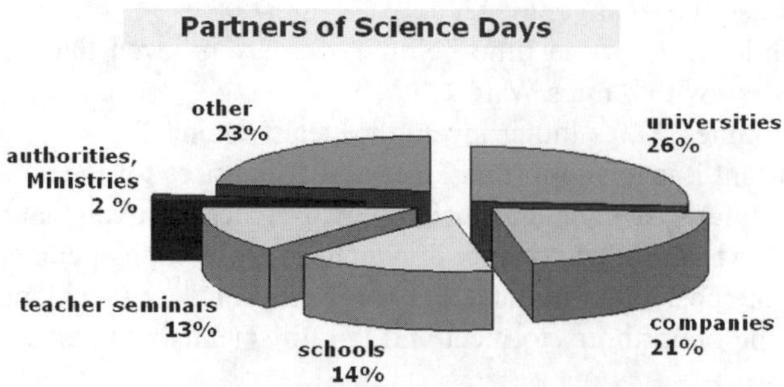

Figure 1. Type of organisations participating in the Science Days event

4. THE PEDAGOGIC CONCEPT

The event's consistent use of pedagogic principles is the reason why the Science Days event has been a success for many years:

- Diversity of activity on offer: various forms of action, topics and partners
- Each visitor has its own specific activity: producing or constructing something at workshops, quizzes, conducting hands-on experiments, etc.
- Learning with all the senses: watching, listening, smelling, feeling, touching, etc.
- Cutting down on difficult subject matter: offer orientation not according to scientific demand but rather on the basis of a prior assessment of the possible needs and knowledge of the target groups.

5. CONSULTATION AND COACHING

All Science Days' partners are entitled to consult the *Foerderverein* on any of the pedagogic or didactic issues. Before an institute's initial participation, an extensive discussion will clarify what the institution would like to offer during the Science Days event. The experience acquired by the *Foerderverein* staff will provide a helpful feedback and influence in the choice of topics and designs. Truly successful formats have often been created in such mutual sessions. The consultation focuses on the following aspects: the definition and orientation towards suitable target groups, the choice of topics, experiments, cooperation with other partners, development of learning and playing materials, development of learning courses, booth design, etc.

6. CARE

We place great value on the participants' well-being during the event. We would thus like to convey to partners that their work is significant for us. Extras such as free meals, beverages, free entry to the Europa-Park, as well as an invitation to the opening programme, the forwarding of the final CD with photos, films, printed articles and the evaluation of the event are always kindly regarded by the partners. In addition, the *Foerderverein* maintains a personal contact with partners even in between events.

7. SUPPORT FOR THE NETWORK FORMATION

At the time of the first Science Days, contacts between the individual partners and the *Foerderverein* were the main concern. However, over the past few years, an intensive network has been built up with more than 200 institutes and in which the different partners communicate and cooperate with each other. Very often this cooperation has been initiated by the *Foerderverein* itself. The most astonishing projects have thus been created, for example:

- "Bionik", a project between the Freiburg University with a grade school
- "Food", a project by the supermarket chain EDEKA and a secondary school

– "Rainforest House", a project with the participation of the WWF and, the Europa-Park Forest Zoology Institute of the Freiburg University and an Educator Seminar in Freiburg

8. QUALITY SAFEGUARDING

In order to maintain the high quality of the event, the organisers place great value on the following matters:

- The goals of the event must be clear to all participants.
- The offers, which have to correspond totally to the concept of Science Days, are selected via a process in which contents and didactic are evaluated.
- The participating institutes must know and consider the mentioned didactic principles (variety of offers, own experiments, reduction of difficult facts and learning with all senses).
- Upon request, experienced staff of the *Foerderverein* will assist individual institutes during the preparation stage and will give consultations on the actual rendering of the theme.
- Because the organizers are present at the venue, an immediate feedback is available if desired.
- An evaluation has been carefully thought through and prepared. Based on this, any exhibitor or organiser will be able to find out the opinion of any visitor who completes the evaluation. This input will be used to improve forthcoming events.
- The organisation must be of a highly professional level, otherwise exhibitors as well as visitors could easily lose interest in the event.

CHAPTER 21

THE CHALLENGE OF SHOWING AND DISCUSSING THE UNKNOWN

NOYURI MIMA

National Museum of Emerging Science and Innovation (Miraikan), 2-41 Aomi, Koto-ku, Tokyo 135-0064, Japan, Tel +81-3-3570-9159, Email n-mima@miraikan.jst.go.jp

Abstract: Over the past 50 years science and technology have evolved at an unprecedented speed. This has resulted in the public being overwhelmed by science and thus confusion and disinterest have spread. The National Museum of Emerging Science and Innovation (Miraikan) was established four years ago in Japan in order to provide a bridge between what seems to the public the bewildering world of research, science and technology, and people. Science museums traditionally focus on public understanding of science (PUS) which presents established science knowledge and examines applications of science in the past. Compared to PUS, Miraikan focuses on public understanding of research (PUR) which examines research into the unknown. PUR must be an on-going presentation following research as it develops. The future impact and applications of research are discussed, presenting the process, including negative and positive aspects, of research as it is happening. It enables the public to engage in a dialogue with the scientists, and to give feedback to the latter. Miraikan assumes an important role as the basis for PUR activities in Japan. For instance, it organizes special 'live talk' events where young and upcoming scientists get a chance to talk about their work to undergraduate students. Such activities promote interaction between scientists/ researchers and the general public and arouse social interest in science and technology

Keywords: PUR, Live talk, Science museum, Cutting edge science and technology, Science communication

1. BACKGROUND

The National Museum of Emerging Science and Innovation (Miraikan) opened in July 2001. It was born as a centre for deepening the understanding of science and technology, and to fulfil Japan's aim of becoming a scientifically and technologically creative nation.

M. Claessens (ed.), Communicating European Research 2005, 119–125.

It was believed to be a matter of extreme importance for Japan to recognize Science and Technology (S&T) as the key for achieving further economic and social development and for sustaining or improving the quality of life in this country. In view of the limited natural resources, rapidly ageing population and poor state of R&D environments, the only way for Japan to cope with ever-intensifying international economic competition and to sustain high industrial standards and quality of life is to create new industries through development of innovative and original advanced technologies.

Under such circumstances, in November 1995, the Basic Law on Science and Technology was enacted with the purpose of helping Japan become a scientifically and technologically creative nation by promoting the advancement of science and technology. In July 1996 the Basic Plan for Science and Technology was formulated to develop comprehensive and strategic plans. The Japan Science and Technology Corporation[1] then decided to create a facility for providing scientific and technological information through exhibitions of the latest science and technology developments, new exhibition techniques, exchanges among researchers and the like. The result is Miraikan.

Although the OECD Programme for International Student Assessment in 2003 showed that Japan is still ahead in mathematical literacy and scientific literacy by international comparison, the number of Japanese students who study science and technology has been decreasing (OECD, 2005). It is predicted that the number of students will further decrease at a relatively rapid pace (figure 1).

Moreover, Japanese students are becoming less interested in S&T fields internationally, which consequently leads to the lack of human resource in S&T fields (figure 2).

In addition, a survey from Japan's Education and Science ministry shows that the public is losing interest in science and technology these days (figure 3).

This has been pointed out also by the science journal *Nature*. "According to the survey, which was released on 10 April [2004], the number of people interested in what scientists have to say has been declining. In 1998, most of those who said they would not attend public lectures on science said it was because they found them difficult to follow. In the latest poll, only one-third objected to the difficulty level – but almost 60% said either that they had no interest or that science did not seem relevant to their lives" (Cyranoski, 2004).

Given these circumstances, the role of Miraikan has become increasingly important. The missions include not just to provide the ideal science education for the children, but also to arouse the interests of the general public as well as to educate the young scientists.

[1] The Japan Science and Technology Corporation (JST) is the funding agency which is focused on research and development from basic scientific research to commercialization with particular emphasis on the creation of new technological seeds. The operating funds are about 11 billion yen (850 million euro) a year.

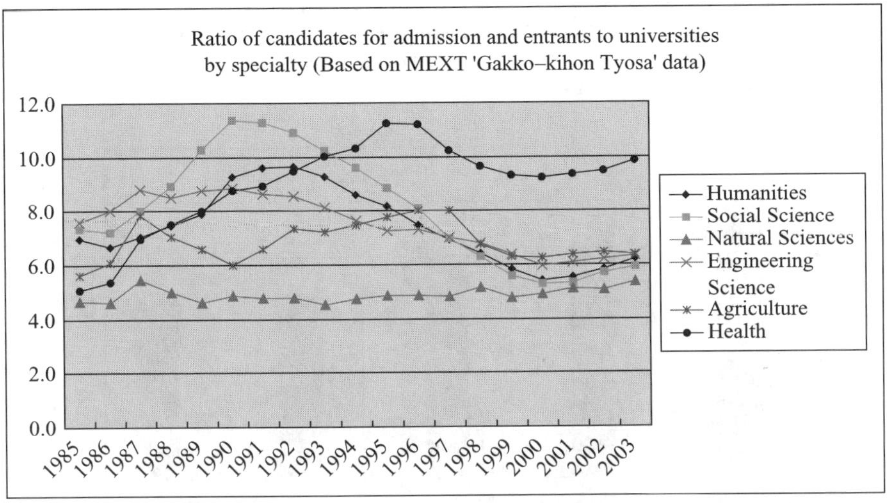

Figure 1. Change in the number of candidates to universities (Ministry of Education, 2004)

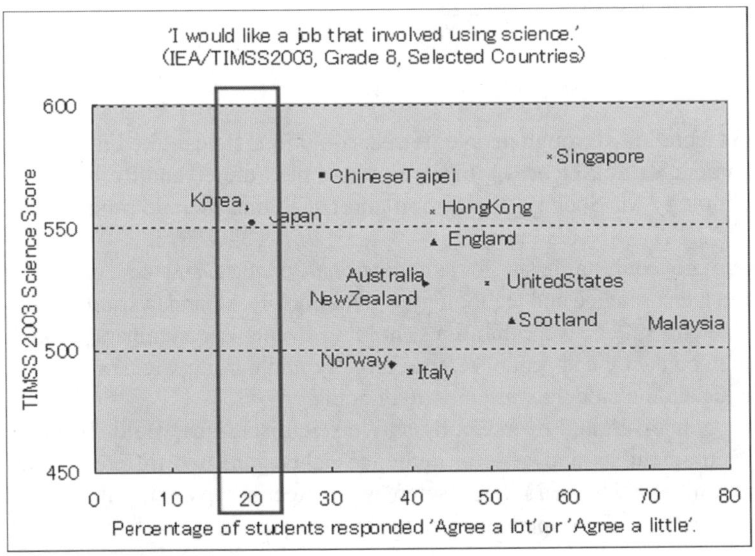

Figure 2. Interest in S&T Jobs (IEA/TIMSS, 2005)

2. SCIENCE COMMUNICATORS AT MIRAIKAN

In Miraikan there are over 200 permanent exhibits in 40,000 square meters. It has approximately 650,000 visitors per year.

The majority of science museum exhibit general science and its phenomena with hands-on or interactive style exhibits. Miraikan is different. It exhibits only cutting

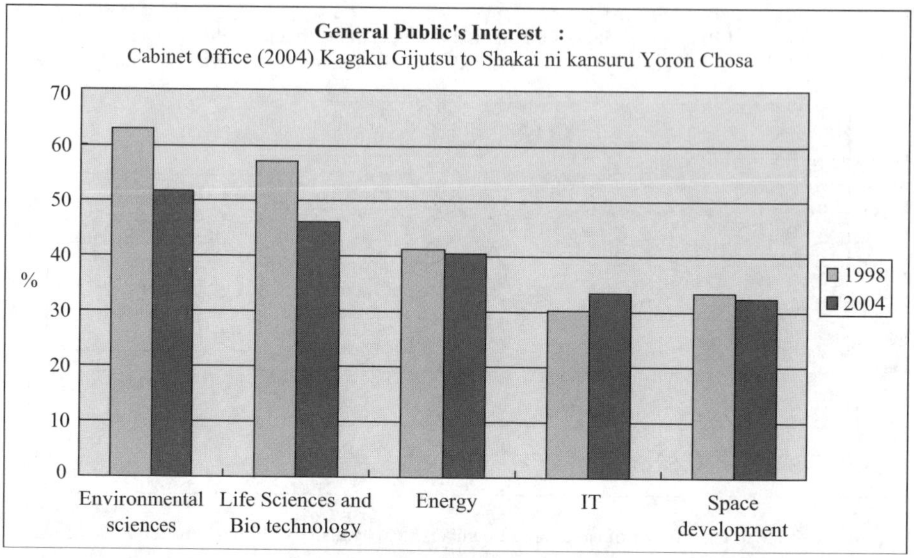

Figure 3. General Public's Interest (Cabinet Office, 2004)

edge science and technology through people. It treats only four fields: 'the Earth Environment and Frontiers,' 'Life Sciences,' 'Innovation and the Future,' and 'Information Science and Technology for Society.' Innovation and the Future include robotics and nano-sciences.

Miraikan offers not only a basis for interpersonal contact and for providing information on the latest science and technology, it dynamically expands various activities, such as producing interactive exhibitions, holding events and seminars through which everyone can contact the appeal of science directly, and introducing new content, etc. Activities inside and outside the museum are ongoing.

Each exhibit is conceived and overseen by active scientists in the field. To make the bridge between scientists and general audiences, there are two different types of science communicators in Miraikan: 'science and technology specialists' and 'interpreters.' There are 80 S&T specialists and 60 interpreters:

2.1. Science and Technology Specialists

We refer to our staff members who are able to interact between the latest science and technology and general citizens and disseminating information on science and technology in an understandable way, as 'science and technology specialists.' They are involved in various activities, starting with the holding of seminars and special exhibitions, planning and public relations activities for exhibits, publications and the website, etc., and have the further role of spreading the methods which were developed here throughout society.

2.2. Interpreters

'Interpreters' are those who give explanations as they translate the latest science and technology into understandable terms. As they explain the exhibits and conduct experiments, they introduce the attractive world of scientists and engineers and the fun of science and technology. Many interpreters and volunteers communicate with visitors at the exhibition space. The exhibits at Miraikan are introduced through 'people.'

3. MIRAIKAN'S MISSION AND SCIENTISTS' INVOLVEMENT

Miraikan's mission is:
- Sharing scientific creativity and activities among people,
- Positioning 'science and technology' as a part of culture, and
- Focusing on creating dialogue between scientists and the community.

To fulfil the mission, we are encouraging scientists to get involved in our activities, such as developing new exhibits together, having science laboratory workshops, holding live-talk events, and operating research laboratory tours.

We invite scientists to Miraikan, collaborate between scientists and science communicators for developing exhibits, and create open and accessible research hubs within Miraikan.

4. LIVE TALK: SCIENCE EDGE

'Science Edge' is one of the new challenge programmes. 'Science Edge' is a live talk event which we hold six times a year. The topics we cover are: brain science, molecular biology, nano-science, planetary science and so on. Target participants are not the general public but S&T major undergraduates. They are not 'audiences' but 'participants.' The guest 'researcher' is neither a 'speaker' nor a 'lecturer.' We particularly care about the word "usage" because we try to promote communication instead of one-way lecture.

We purposely select young upcoming scientists as guests because we want to change the stereotypical image of a scientist: white lab coat, messed up hair and thick eyeglasses. We select 'hot' and 'cool' researchers from post-doc or doctoral course students. "Hot" means passionate, enthusiastic about their research topic and "cool" means smart, fashionable, up-to-date young researchers. We like to introduce the researchers as they are real people to the public.

Each event is associated to a host, who plays an important role (figure 4). The host is a science communicator in Miraikan who is a Ph.D. holder or who has equivalent knowledge and skills. She or he:

1. Investigates in most recent academic journals such as 'Science' or 'Nature' to select a guest researcher,
2. Interviews the researcher to understand their passion, and

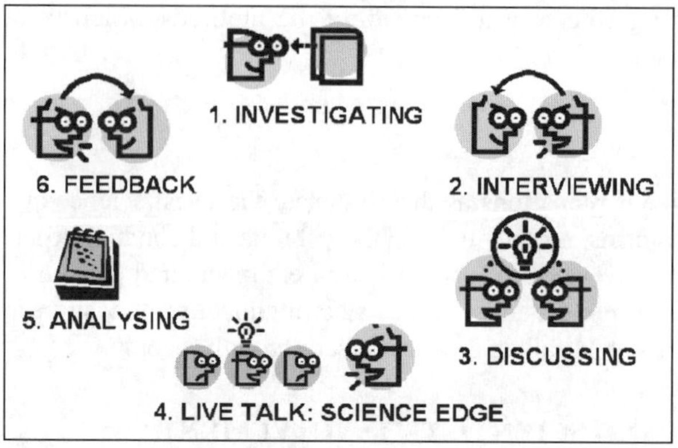

Figure 4. Roles of the host at 'Science Edge'

3. Discusses a theme and logistics for a Science Edge event.

The key is

4. To have a talk with the researcher, the host, and undergraduate students after a talk,
5. To distribute and collect questionnaire sheets from participants and analyze the result.

Then we provide

6. Feedbacks to 'the guest researcher' to improve their communication skills.

5. MAKING BRIDGES BETWEEN DIFFERENT COMMUNITIES

We are trying to make a bridge between different communities, especially between the public and the S&T community, but also within the public and within the S&T community. We link science communicators to the public and to S&T community in a variety of ways. One of them has been Science Edge.

We are trying to connect to the public in different ways; not only through visitors, but also through researchers, through mass media, through volunteers, through industry, through government, through schools, and through other science museums. And we link the researchers with all of the other groups. We have just taken the initial steps. We are trying to show the underlying process of science and scientists as it is happening. We believe this will attract young people. As we better our approach, we bring the wonderful world of science and technology to more and more people

BIBLIOGRAPHY

Cabinet Office, Government of Japan (2004) Kagaku Gijutsu to Shakai ni kansuru Yoron Chosa. (Opinion Survey on Science & Technology and Society)

Cyranoski D (2004) Japanese nobels fail to inspire interest in science. Nature 428:787

International Association for the Evaluation of Educational Achievement (IEA) / Trends in International Mathematics and Science Study (TIMSS) (2005) Mathematics and Science Achievement of Eighth-Graders in 2003

Ministry of Education, Culture, Sports, Science and Technology (2004) Gakko-kihon Tyosa data (Basic Survey on School)

OECD The Programme for International Student Assessment (PISA) (2005) Learning for Tomorrow's World – First Results from PISA 2003

CHAPTER 22

SCIENCE & THE CITY

A metropolitan strategy to promote scientific culture at local level

VLADIMIR DE SEMIR

Commissioner for Scientific Culture, Barcelona City Council, Science Communication Observatory, University Pompeu Fabra, Paseo Circuncvalacion 8, 08003 Barcelona, Spain, Tel: +34-932-689262, Email: vladimir.semir@upf.edu

Abstract: The *Science & the City* programme aims to disseminate scientific culture in the city of Barcelona (Spain). It is integrated within the framework of the Strategic Plan for the Cultural Sector of Barcelona and includes a set of short, medium and long-term actions, which are carried out throughout the city: in districts, libraries, civic centres, schools, hospitals, universities, etc. Barcelona proposes to build up a European network of cities (ESCITY) to work together and share experiences in relationship with the dissemination of scientific culture. ESCITY focuses on creating the core for a community that brings together regional and local governments and social researchers across Europe to exchange information and best practices with the emphasis on strategies that place the promotion of scientific culture within the cities' and regions' cultural policies

Keywords: Scientific culture, Local political initiatives, Science communication, Public understanding of science, European network of science cities

In July 2002, all the political groups represented in the Barcelona City Council voiced their unanimous support of the "Science and Society" action plan drawn up by the European Commission. This commitment to Europe was a decisive step for incorporating scientific culture into the different areas of municipal government, particularly in the Culture Institute of Barcelona. The promotion of scientific culture is, today, a part of the political agenda of the local government of the city of Barcelona. This was reaffirmed in November 2004 with the presentation, in a plenary session of the City Council, of the strategic plan *Science & the City* for the period 2005–2015.

The *Science & the City* programme constitutes the drive of the Culture Institute of Barcelona, in close co-operation with the rest of municipal areas and departments

M. Claessens (ed.), Communicating European Research 2005, 127–132.

to promote scientific culture among citizens. As well as receiving strong political support from the areas responsible for its implementation, this strategy is supported by analysis and monitoring carried out by the Science Communication Observatory at the University Pompeu Fabra. At the initiative of the Culture Institute of Barcelona, this university will become the location of the future Academy of Society & Science. The Academy will be a think-tank born as a legacy of the Universal Forum of Cultures – Barcelona 2004 event and an initiative derived from the international network Public Communication of Science and Technology (PCST).

In the coming years, the PCST Network will experience an important expansion of its global influence, but we have to work to guarantee that the activities of those countries with more financial difficulties will be incorporated into the network. It is not only about the PCST, a network which is continuingly growing, but it is related to a wider representation of the different cultures in the world. In this sense, science communication has to respect different cultural contexts and integrate knowledge from all the continents. This is one of the main PCST targets. This present and future cultural diversity is the greatest PCST heritage!

This is another relevant legacy from the Universal Forum of Cultures – Barcelona 2004 related with the spread of scientific culture and the commitment of the Barcelona City Council to do this at the local political level.

The Local Authorities Forum of Porto Alegre – mayors and local government representatives from all over the world committed to social inclusion – met in Barcelona in May 2004 and proclaimed the basis document named Cultural Agenda 21. The idea arose to contribute formulating answers to the cultural development challenges we are now facing in the 21st century, given that the tendency of exclusive, uniform globalisation threatens culture.

The initiative has a clear precedent in the process toward the end of the last century regarding the environment, when it was evident that, current development models, which are excessively dependent on resources and natural ecosystems, first mobilised world public opinion which later irreversibly influenced government and international institutions. There are indisputable political analogies between cultural and ecological problem areas given that culture as well as environment is the common wealth of humanity. Today, a similar sensitivity is emerging in the cultural field, for this it plays a central role in globalisation but there are not enough instruments for a true political debate.

Thus it is especially important to develop agreements to define the cultural diversity and preserve the values of creativity, participation and inherent freedom of the cultural generation. At the same time, it is also important to influence the necessary balance between knowledge and power. In this sense a new sustainable development is already on the scene related to the binomial: knowledge and power. These are the fundamentals carried out by the Local Authorities Forum that led Barcelona and Porto Alegre to suggest Cultural Agenda 21.

Local governments are currently working on a key document to propose the need to boost the open and diverse cultures in a world characterised by globalisation. It is also important to adopt the commitment for culture to have an essential dimension in urban politics and in real participatory democracies that support social cohesion. A relevant aspect of the Cultural Agenda 21 is the compromise to "encourage programmes directed at science dissemination and technological cultures among all citizens". They consider that the application of new scientific knowledge generates ethical, social, economical and political questions that are of clear public interest.

The *Science & the City* programme of Barcelona works in this direction, with the will to share experiences and build networks of co-operation with other cities throughout the world, because sharing scientific culture is also a universal process.

1. THE CHALLENGES

Barcelona is a city with a solid scientific and technological tradition. A large part of this tradition is linked to the considerable development which took place following the industrial revolution and which gave prominence to sectors such as engineering and biomedicine. But, as well as these traditional values, the transformations which derived from the expansion of the knowledge society have led to the emergence in recent years of new sectors in the city and there are even more changes to be undertaken.

Science & the City is Barcelona's strategic plan for promoting and disseminating scientific culture, the traditional values of the city and the new strategic sectors. In addition to all that, it is a response to the problems and shortcomings which already exist and which are at risk of becoming worse as the process of the expansion of science and technology moves forward. Specifically, the *Science & the City* action plan presents the following 10 challenges for the future:

1. To encourage a more active population with greater decision-making capacity in the face of new scientific, medical and technological challenges;
2. To consolidate the public image of research and innovation as activities which generate wealth and, therefore, as key elements of development;
3. To improve the chances of accessing new opportunities, thus strengthening social cohesion and reducing the educational and digital divides;
4. To drive the transformation process of Barcelona's science centres and museums towards the new Museum of Natural History of Catalonia – our region – and, at the same time, promote the role of these elements, which articulate the scientific culture of the city;
5. To strengthen the city-university-enterprise triangle by encouraging synergies between the three sectors;
6. To identify and promote the key strategic areas for the development of Barcelona in the knowledge society;
7. To encourage knowledge and dissemination of local scientific values coming out of the university environment;
8. To redirect the current crisis in scientific vocations among young people, guaranteeing the future competitiveness of the city in terms of research and development;

9. To correct the gender imbalance in the area of science and technology by encouraging greater participation and visibility of women;
10. To promote the values of Barcelona internationality as an innovative city which generates ideas, projects and opportunities.

2. THE PROGRAMME

Science & the City is integrated within the framework of the Strategic Plan for the Cultural Sector of Barcelona and includes a set of short, medium and long-term actions, which are carried out throughout the city: in districts, libraries, civic centres, schools, hospitals, universities, etc. Proximity to the people is precisely one of the elements which local government can provide; in this case, Barcelona City Council. In addition to this proximity, there are multiple formats, some traditional and other innovative, but all thorough in their implementation and follow-up.

Apart from these actions, *Science & the City* is included in the specific annual programmes that define the city's cultural agenda. Thus, for example, 2005 was the Year of the Book and Reading throughout Spain at the initiative of the city of Barcelona, meeting with the fourth centenary of the publication of *Don Quixote*. In this context, a symposium was held in November 2005 on the role of popular science books and the new opportunities provided by current technologies in order to encourage reading in the area of science – often the source of scientific vocations among young people. At the same time, an annual fair of popular science books was created. Another example of these annual programmes is the aim to declare 2007 the *Year of Science*, thus reaffirming a process of scientific culture integration in the cultural planning of the city.

As well as the annual programmes, there are other regular or permanent activities. These include the *Science Autumn*, *Women and Science, Science on the Street*, the activities from the development of the *Academy of Society and Science*, the *Barcelona Forum of Innovation and Scientific Culture,* etc.

The *Science & the City* programme is strictly a municipal initiative. It is planned and developed in collaboration with all sectors where the citizenship is represented, with a special emphasis on the role of the universities and research centres of education, enterprises and the industrial sector.

Here are some of the specific actions:

- Cycle of lectures "Genetics and City" aiming to sensitize citizens on the advances in this area of knowledge that opens many doors to hope, but also ethical challenges.
- Popular itinerary through Barcelona and its neighbouring countries (each country will have a stand) at the international exhibit "Science in our pockets" and which highlights significant people in the history of science using notes of money which the various countries used before the euro was introduced.
- Popularization lectures, "Science in the neighbourhood" in the civic centres in the various neighbourhoods to bridge the gap between scientists and Barcelona's popular culture. Airing on the local television channel BTV of the lectures celebrated during the cycle.
- Publication of the tour guide "Walks through scientific Barcelona" that provides the city with a guide to the knowledge of its scientific history and

current state, similar to those guides existing for Paris and London and which present several scientific itineraries through the city for schools and other cultural activities.

- Organization of the series of lectures "Barcelona's scientific look" analyzing Barcelona from geological, biological, chemical, physical, mathematical points of view and that will be aired on the municipal local TV channel, BTV.
- Production, in co-operation with the Science Communication Observatory at the University Pompeu Fabra of 26 "capsules" (short programmes of 5 minutes) for the local channel BTV –funded by the Science and Technology Foundation, of the Science and Technology Department in the framework of the national programme for the popularization of science – in which 26 scientists from Barcelona provide brief summaries of their research and the reasons for which they conduct these studies as part of their scientific vocation.
- Conducting of a quantitative and qualitative study, in collaboration with the Audiovisual Council of Catalonia of the science programmes aired on Spanish and Catalan televisions.
- Deployment of science popularization in high schools in Barcelona in collaboration with the Municipal Institute of Education in Barcelona.
- Organization, with the department of Policies for Women and Education from Barcelona City Council, of the Symposium "Women, Men and Science", which will hold a debate on gender-related aspects involving scientific vocation and careers and promoting equality among women and men in science, a point proclaimed as one of the goals of the "Science and Society" action plan of the European Commission.

Of course, these initiatives are not strictly new, but they are usually led by universities, science centres, labs . . . rather than the local governments of the cities who have a wider area of influence in society. This is the main objective: to extend the programmes and action to more people. The scene is still in a very preliminary stage and, though in the following years progress will certainly be made, we should foster the development of city networks and collaboration among administrations, in order to cooperate and share experiences in the promotion of scientific culture. On the other hand, a part of the current media – particularly local or regional TV channels and radio stations – belong to the local or regional administrations and what clearly promotes the possible impact of these actions, as they are the "proximity media" of the citizens.

Furthermore, we must take into account that such administrations are also responsible for the cultural offer and construction of the cities in the 'conventional' sense of the word (i.e. those responsible for the spread of the humanistic and artistic culture). So, integrating the scientific culture as another type of culture is one of the unresolved matters of the European local administrations.

3. THE NETWORK

The *Science in Society Forum* held in March 2005 in Brussels underlined the need for a greater institutional commitment across Europe in raising the profile of science as an integral part of our culture. Despite the advances made since the launch of the Commission's "Science and Society" action plan, little has been done so far to erode the borders between the "two cultures" and promoting scientific culture is not usually an issue of our cultural agendas. Besides, most actions are implemented at a national level by scientific institutions while the municipality is the administration closer to citizens and, probably, the most flexible in incorporating new public engagement tools.

For this reason, the municipality of Barcelona proposes to build up a European network of cities (ESCITY) to work together and share experiences in relationship with the dissemination of scientific culture. ESCITY focuses on creating the core for a community that brings together regional and local governments and social researchers across Europe to exchange information and best practices (with the emphasis on strategies that place the promotion of scientific culture within the cities' and regions' cultural policies). The project aims to settle the basis for a wider and lasting network, able to act as a lobby to strengthen the promotion of scientific culture as a basic item on the cultural agendas of European cities and regions.

These are the main objectives:

- To provide a broad forum for scientific culture issues across Europe.
- To benchmark the promotion of scientific culture at regional and local levels.
- To organise a European Conference to be held in Barcelona in 2007, which the city has declared "Science Year". Prior to this, contacts will be made with other European networks of cities in order to gather experiences, concerns and suggestions relevant to ESCITY and identify potential participants and network members.
- To engage in a lobby strategy by publishing an action guide to be distributed to European governments, its contents being the forum conclusions, a list of best practices, the benchmarking main findings and a set of tools and recommendations for local and regional governments.

ESCITY is a two year project that started in March 2006. It aims to provide a solid basis for this new community to expand and consolidate beyond the Sixth Framework Programme. The project brings together 6 institutions from 5 EU member states (Spain, Austria, Italy, Portugal and Norway). Among those, 4 are governmental institutions that are currently engaged in the promotion of scientific culture, and the other 2 are research centres working on issues regarding science communication and public understanding of science.

Participants hope that by the end of this project many other European cities will join the ESCITY network and scientific culture will be a part of the political programmes of their local governments!

CHAPTER 23

THE GENOVA SCIENCE FESTIVAL

MANUELA ARATA

National Research Council and Genova Science Festival, C. so Perrone 24, 16152 Genova, Italy, Tel +39-10-6598740, Email m.arata@infm.it

Abstract: The third edition of the Genova Science Festival (Italy, 27 October – 8 November 2005) has concluded with extraordinary results 54 000 sold tickets (between passes and dailies), a strong increase of 50% in respect to 2004 and a total of 216 000 visits to the numerous programmed events. The article reports on the reasons for this success, and launches the idea for a new initiative: a "European Academy for Scientific Explainers"

Keywords: Public awareness of science

The Genova Science Festival was organized for the first time in 2003. It brought together national and local institutions. Although the initiative was welcomed, it also generated some fear. Even if in the past, science exhibitions promoted by the Italian Institute for the Physics of Matter always registered a large attendance, it was impossible to know in advance how the general public would welcome this event given the well-known reluctance of the Genoese people.

On the other hand, the festival was a marvellous dream come true: i.e. to organize a science event that could gather together people from all ages and interests with an innovative formula. Science for everyone and everywhere with twelve days of meetings, science and art exhibitions, educational activities, shows, round tables, book presentations, video projections, games, performances, workshops, and special events, often mixing art, literature, cinema and even gastronomy, alternating experiments with theories, games and performances to widening conferences.

And if in 2003 the Festival at its first test met with great (and rather unexpected) success, the figures of the second and of the recently-concluded third edition strongly confirm the great interest and attention for science from the whole society. The 2005 edition of the Festival (27 October – 8 November) ended with extraordinary results: 54 000 sold tickets (between passes and dailies), a strong increase in

M. Claessens (ed.), Communicating European Research 2005, 133–136.

respect to 2004 (in which ticket sales were 36.000), and a total of 216 000 visits to the numerous programmed events.

The 77 locations, dispersed around Genova and Liguria, hosted 250 events (141 conferences, 37 events, 36 between shows and films, 33 workshops, 28 exhibitions), almost always registered "sold out": a participating public which was informed and prepared, not only prized the more noted names but also the specific themes, discussions and debates.

While the 2004 edition was one of the official highlights of "Genova 2004 European Capital of Culture" and part of the "European Science Week", the 2005 edition paid particular attention to the "WYP – World Year of Physics 2005" and to the "Researchers in Europe Initiative".

Among the exhibitions, physics and education met in "Simple and complex 2005" – which returned more extensive and more interactive-, while the exhibition dedicated to the fossils discovered in Patagonia "The secrets of the dinosaurs" illustrated the latest devices adopted to discover the history and secrets of these fascinating creatures. Art and science were at the centre of the exhibition "The four elements", which spotlighted the way in which contemporary art interprets nature.

An intense programme of shows compared the different languages of science, dance, art, television documentary, theatre and cinema, including the world premiere of "The Children of Uranium" (by Peter Greenaway and Saskia Boddeke), and the Italian premiere of Peter Parnell's play "QED. A day in Richard Feynman's life", by the young Director, Luca Giberti.

Most of the scientists and performers involved came from all over the world and included Amir Aczel, Andrea Ballabio, Giovanni Bignami, Saskia Boddeke, Edoardo Boncinelli, Pep Bou, John Brockman, Marcello Buiatti, Lella Costa, Alvaro de Rujula, Ludovico Einaudi, Richard Ellis, Brian Fagan, Richard Fortey, Owen Gingerich, Giulio Giorello, Pietro Greco, Peter Greenaway, Brian Greene, Robert Laughlin, Andrea Liberovici, Benoit Mandelbrot, Piergiorgio Odifreddi, Giorgio Parisi, Roger Penrose, Carlo Alberto Redi, Martin Rees, Hilary Rose, Steven Rose, Patrizio Roversi, Merritt Ruhlen, Paul Seabright, Gino Segre, Sandro Stringari, Craig Venter, etc.

It was quite extraordinary to see whole families, senior citizens, students, professionals, housewives and intellectuals crowding their conferences, queuing up to reach the conference sites, and debating about evolution, stem cells, or the controversy regarding the great theories of the cosmos . . . It was a strong signal that made us understand the extent to which people want to know about science and culture.

Some scientists – e.g. the mathematician Pier Giorgio Odifreddi – had to repeat their conference in order to accommodate the number of people interested in attending, while the theoretical physicist Gabriele Veneziano had to tell "The theory of strings as a possible realization of Einstein's dream" while standing, because the stage was full of spectators of all ages, who attended the conference with notebook and pen in hand: this was a concrete example of how the Festival eliminates the distance between science and the public!

The media coverage of the Festival was also excellent: the most important media was represented by 244 accredited journalists, and there were 30 television broadcasts (national, satellite and local), 26 radio broadcasts, and hundreds of articles published in newspapers as well as magazines during the event.

The press office of the Festival issued 72 press releases in coverage of the principal events, and the web site www.festivalscienza.it registered 1 250 000 hits from September to November 2005.

There was a notable increase in attendance from schools, often from outside Genova and Liguria. 31 410 tickets were sold (2 570 from Lombardia, 1 531 from Piemonte, 1 523 from Toscana, 422 from Emilia Romagna, 239 from Lazio and 170 from Veneto). Young visitors, in particular, demonstrated a feeling for science and the wonders that it can offer. Beyond the practical workshops, where they could – for instance – touch and play with different materials and experiment with the laws of physics, the children applauded enthusiastically the soap bubbles of Bufaplanetes (Pep Bou), as well as the explosions of light from the Chemical Wild West and High Energy (Andreas Korn-Mueller). Particularly appreciated were the workshops "Focus Bio Bubble", which made it possible to enter into a cell enlarged 450 000 times, and "Scientific Police at work", which involved the children in discovering the culprit of a "murder" by allowing them to use state-of-the-art techniques on ballistics, fingerprints, DNA and drugs!

A very "special job" was carried out annually by the over 400 young entertainers of the Festival, responsible for involving the public and explaining scientific exhibits. They were selected after a public call and were chosen for the various initiatives based on their attitude. After a common training on general issues such as the diffusion of science, organisation, the relationship with the public and communication and presentation skills, they were sent to the group which was responsible for preparing exhibitions and events for education on specific topics. In 2004, the group was integrated with students and researchers from France, Slovenia, UK, and Spain who were participating in the "Esciential – European Science Festival" Project funded by EU and with about 20 students from Europe and the Far East within the Aesotope Project (ASEF-INFM initiative).

Many of the teenagers coming from non-scientific disciplines were attracted to science by this fantastic and exciting experience and many of the scientific graduates decided to enter a research career thanks to their involvement in the Festival!

The Festival showed clearly that scientists are not prejudiced at all towards science diffusion and education, but they often need to be "educated" in showing the right attitude to prove this.

Universities and research institutions still do not provide any training for "edutainment" or "animation" for researchers, which is surprising because it is essential for future researchers to be able to communicate their latest advancements directly to the public, whether they are journalists, school teachers, students and/and citizens. As science diffusion and education should be considered as a "normal" duty, it is crucial to provide researchers with methods and instruments to ensure good interaction with the general public, and to create new professionals.

Although this has already been launched at a European level, it is certain that the time is right to start a sort of "European Academy for Scientific Explainers" (EASE), offering courses within scientific PhDs, Masters, short courses and schools for high-school students, stages and mobility among science centres and museums, universities, festivals, etc. aiming at creating a new sensitivity among the next generations towards science and bringing new teams of people who speak the same language on the "research market".

This is the reason why in October 2006 the Genova Science Festival will organize the first EASE Summer School, which will provide interdisciplinary education overcoming the traditional disciplines, and will focus on the new frontiers of science: nanosciences, bio-informatics, quantum technologies, genomics, environmental sciences, etc.

The School will provide students with communication techniques, theatre and body management, entertainment, organisation, design and construction of exhibits and multimedia, and also provide the opportunity to get practical experience, such as preparing exhibitions and workshops, as well as "on site training" during the Festival itself. In this sense, the School will be open to participants coming from all European Countries and we are seeking support within the European Union's research Framework Programme to cover part of the costs for participation of PhDs and post-docs.

It is hoped that the Science Festival, which will spread in other Italian cities by the end of 2006, becomes a spark capable of lighting up many more events, of attracting and moving people from all over Europe, and of finding new places and opportunities where science can be promoted. As internationalisation is the key for success, this event could become a recognised European reference point for the popularisation of scientific culture at all levels.

The 2006 Festival will be held from October 26 to November 7 and the keyword will be "Discovery". European colleagues are most invited to participate and propose events and exhibitions.

CHAPTER 24

SCIENCE CLASS 2012

A vision of classroom science teaching in the future

RUSS HODGE

European Molecular Biology Laboratory, Meyerhofstrasse 1, 69117 Heidelberg, Germany, Tel +49-6221-387252, E-mail hodge@embl.de

Abstract: Dramatically improving science teaching throughout Europe will require formulating a precise, practical vision of the end goal: the science class of the future. This article provides a recipe for what that class might be like and how to assemble the ingredients: new types of mechanisms to enhance the flow of information between the worlds of research and the classroom, to give scientists practical experience in communication and teaching, to train teachers in new ways, and to create interdisciplinary science teaching that is linked to other disciplines, such as languages and ethics. EMBL and the EIROforum have developed pilot projects to try out some of the techniques that are proposed, and our experiences are briefly described

Keywords: Teacher training, Research and education, Science communication, Science teaching, Communication training, Science teaching journal

Let us step into a science class somewhere in Europe of the year 2012, first day of the school year. There are new books on the tables, reflecting a spiced-up curriculum. After two years of thrifty saving, the school has invested in a large-screen monitor mounted at the front of the room; it will be used for teamwork with other schools, for virtual visits to laboratories and to communicate with a submarine searching for new forms of life on the ocean floor. The teacher, Ms. Keynes, has just turned in her lab coat in exchange for a grade book. After seven years of studying how dolphin genes respond to changes in the environment, she wanted a change of scene, and now has a two-year contract with the school. After that she may stay, or she might decide to return to the lab. Alongside her teaching duties, she is helping her colleagues catch up with modern biology.

A handout describes the plan for the semester. The first surprise is that the class will be held in English – "sorry, but with English you can work in science anywhere, and you'll need it when you give conferences in exotic places." Each student should

M. Claessens (ed.), Communicating European Research 2005, 137–141.

pick a scientific journal and read it regularly. The class will keep a scrapbook of science in the news. Virtual excursions to labs will be followed up by visits from practising researchers – "some of you have never met a scientist; they're normal people, and they can tell you what a career in science is really like." Every two weeks, the physics teacher will stop by to talk about X-rays, optics and biophysics, and the computer lab will introduce bioinformatics. The process of science will be as important in the course as learning facts. With their ethics teacher, the students will make posters about upcoming scientific themes, and discuss decisions they will likely have to make in the future.

In 2012, classes like this are being conducted all over Europe, made possible by instruments put into place as part of plans for the European Research Area (ERA), and hard work by many stakeholders. Now the ERA has been largely achieved, and a larger European Union (EU) is regularly fed innovations from fundamental and applied science. Schools are expected to do their part in preparing students for the future – whether they study science in college, or simply are to become scientifically literate citizens.

*****

This could be the science class we have in a few years, but getting it will require developing a very clear picture of what it should be like, and then working backwards from the vision to create the necessary educational, political and other tools necessary to build it. The alternative is to try to "cure" problems in current educational systems by addressing single symptoms one-by-one. That can work, but it will mean a wide range of quality in teaching across Europe, and without an extremely clear vision of the goal, it will be difficult to measure progress. The starting point will vary widely from state to state, nation to nation, but the ultimate goal should be nearly the same.

Several factors make this a European issue. Science and science careers are thoroughly international, and teachers everywhere face similar problems in preparing their pupils for the modern world. The pace of science is so rapid that school teachers in science will need to be trained in a different way than their colleagues in other fields; there are some recipes for doing so, and the wheel should not be reinvented at the local level. Finally, training scientists for the next decade will require a new type of collaboration among research institutions, universities, teacher training colleges, schools, and many others. This will have to be tried out in a few places, as pilot projects, and then quickly extended as widely as possible. The impetus to do so will have to come from a much higher level than state ministries.

This article can only begin to list the ingredients of the science class of 2012 and make a few suggestions about where to shop for them. The picture needs to be fleshed out in detail and implemented. The points here are derived from discussions with teachers, scientists, students, and my own children – end users who typically come home with a very disappointing report on the state of science education. Some of the mechanisms for moving forward are already in place, often directly due to support from the EU, and we believe that the strategy described in the European Commission's Science and Society Action Plan is largely the right way to continue.

The first goal should be to design a class that excites students about science. While surveys show that the European public is generally interested in science, students continue to describe their science classes as "boring". This is almost incomprehensible to anyone who knows what is happening in research. For example, a *single week's* issues of the journals Science and Nature contained these stories:

- Enormous explosions in deep space allow scientists to measure vast distances in the universe
- A study of Amazon tribes suggests that there may be universal concepts of geometry among humans – some of these may be innate, rather than products of culture or any specific learning environment
- Some species of bacteria living near the North Pole sense the Earth's magnetic field and swim north; others swim south, probably by detecting changes in gradients of oxygen in water
- Simulations show that a few million years ago, when the rotational axis of Mars was tilted 45 degrees from its current position, large glaciers may have existed on the planet's surface
- Scientists conducted large-scale sequencing of DNA extracted from the fossil remains of mammoths
- Dust from the ocean floor tells us about a collision that occurred between two asteroids – one of which was over 150km in diameter – about 8.5 million years ago
- Studies of a fossil fish reveal that the ears of fish, birds and mammals may have originally evolved from parts of the respiratory system

Such stories showing where research is and where it is going are usually more interesting than the news of twenty years ago, when most school teachers learned about science. Many of the things on this list fit into current curricula and could spice it up – if teachers heard of the stories and could understand them. This does not happen nearly often enough, because of wide gaps between schools and research – and between schools and university education.

Some of the mechanisms needed to keep teachers up-to-date in an era of rapid discovery are obvious:

- A regular flow of comprehensible information from the research into education communities, through talks, websites, journals, science ambassador programmes, etc.
- Future-oriented training in teacher education programs, with much more time in the lab and direct contact with scientists
- More training in teaching the process of science
- Access for teachers and pupils to major scientific journals (often too expensive for schools or individuals to afford)
- Effective, practical continuing education to update teachers regularly on progress in subjects they are teaching (and parallel fields)
- More interdisciplinary training and experience in areas that are coming much closer together, such as biology, physics and chemistry
- Mechanisms to encourage scientists to become teachers, and more opportunities for teachers to participate in real scientific studies
- More interactions between disciplines and other courses which cover, for example, ethics and language skills

- Enough flexibility in the curricula to make room for current events

Research institutions like the European Molecular Biology Laboratory (EMBL), the other major international European research organisations (EIROforum[1]) and universities and institutes at the national level have an important role to play in this process by:

- Providing high-quality information about their work, making teaching kits and classroom projects based on cutting-edge science
- Offering workshops and summer schools to teachers and students
- Encouraging their scientists to visit schools and take part in other educational events
- Defining communication and didactics as core skills for research careers, and including them in education programmes for young scientists; this will be helpful in their scientific careers and might encourage them to become teachers
- Encouraging scientists to talk to the public about the social, ethical and applied dimensions of their work
- Providing other types of platforms for interactions between the worlds of research and education

These are all areas in which the European Commission, national funding agencies, private foundations and others have made a profound impact and continue to do so. External funding is usually essential in allowing research institutions to become involved in a significant way, because it is often difficult for them to direct funds away from their primary mission (research).

What has happened at EMBL is a good case study of how support can be channelled into meaningful, long-term projects. A grant from the EU permitted the creation of the European Learning Laboratory for the Life Sciences (ELLS).[2] Once in place, its value to EMBL was recognized, and the Laboratory decided to provide core funding for it to continue. This facility has various functions:

- It regularly brings in international groups of teachers for workshops on modern topics from molecular biology
- It uses these workshops as platforms in which scientists get the chance to communicate and teach
- It helps EMBL scientists turn their work into teaching materials and lessons
- It is helping to spawn a network of like-minded initiatives
- Staff travel to other countries and institutes, helping them launch similar programmes

Another project hosted at EMBL and run by the EIROforum is an international journal for science teachers called *Science in School*.[3] It will be launched in spring 2006, collecting excellent science teaching projects and passing them on to other teachers. The journal will also contain news from current science, articles about events and workshops, profiles of scientists and teachers, etc. The journal is part of a much larger EU-funded cluster called Nucleus,[4] which includes other activities aimed at teachers. One of these is Science on Stage, a sort of "Olympics for science teachers"

---

[1] www.eiroforum.org

[2] The European Learning Laboratory for the Life Sciences: www.embl.de/ells

[3] www.scienceinschool.org

that consists of national events in most European countries and an international festival. Science on Stage has demonstrated that Europe has brilliant science teachers already doing excellent work – sometimes in spite of considerable constraints. They find working with international colleagues extremely valuable, although education policies and practice are usually determined at the national or state level.

Events and mechanisms for exchange will be very helpful, but the ultimate goal should be to create a culture of science communication to buttress the European Research Area. Excellent science teaching in our schools is a prerequisite to achieving this. Alongside everything else that has been done, we must develop an absolutely clear picture of the science class of the future. The destinies of our institutes are intimately tied to both the quality of Europe's science students and the success of the ERA. The EIROforum organizations are doing what we can to stimulate this process from our side, but it must be done in partnership with teachers and schools, museums and science centres, media and communications experts, policymakers at several levels, and all other stakeholders in science education. Nucleus and our other education projects are a step along the way, and we invite others to join us through mechanisms such as the journal and Science on Stage.

[4] www.xplora.org

CHAPTER 25

SCIENTIFIC LITERACY

Can the school science curriculum deliver?

ROBIN MILLAR

University of York, Department of Educational Studies, Heslington, York YO10 5DD, United Kingdom, Tel +44-1904-433469, Email rhm1@york.ac.uk

Abstract: 'Scientific literacy' is increasingly seen as the primary goal of school science. To address it, we must first acknowledge the tension within the school science curriculum between the aim of enhancing all students' scientific literacy, and of providing a sound foundation for more advanced study in science for some students. The *Twenty First Century Science* course shows one way in which this can be done. A large-scale pilot is indicating that many teachers welcome a curriculum model that separates these two purposes more clearly, and a course that focuses on scientific literacy – and report that the response and engagement of their students is significantly improved

Keywords: Scientific literacy, School science, Curriculum development

1. INTRODUCTION

In recent years the term 'scientific literacy' has become increasingly prominent in discussions of the aims and purposes of school science education. The 1990 UNESCO World Conference on Education for All argued that science education should promote 'a world community of scientifically and technologically literate citizens' (UNESCO, 1990; Layton et al., 1994). The US *National Science Education Standards* claim to 'present a vision of a scientifically literature populace [and to] outline what students need to know, understand, and be able to do to be scientifically literate at different grade levels' (National Research Council, 1996: 2). In the UK, the *Beyond 2000* report (Millar and Osborne, 1998) argued that 'the science curriculum from 5 to 16 should be seen primarily as a course to enhance scientific literacy' (p. 9). In Australia, a major research report for the Department of Education, Training and Youth Affairs

M. Claessens (ed.), Communicating European Research 2005, 143–148.

(Goodrum, Hackling and Rennie, 2001) starts from the premise that 'the purpose of science education is to develop scientific literacy which is a high priority for all citizens' (p. ix). Reports from other countries express similar sentiments. Perhaps the clearest indication of the prominence of 'scientific literacy' in current thinking about the science curriculum is the decision of the Organisation for Economic Cooperation and Development (OECD), in their Programme for International Student Assessment (PISA), to survey students' 'scientific literacy' rather than their understanding of school science (OECD, 1999; Harlen, 2001).

This principal reason for the growing concern about scientific literacy is clearly expressed in the European Commission (1995) *White Paper on Education and Training*. It argues that:

> Democracy functions by majority decision on major issues which, because of their complexity, require an increasing amount of background knowledge. . . . At the moment, decisions in this area are all too often based on subjective and emotional criteria, the majority lacking the general knowledge to make an informed choice. Clearly this does not mean turning everyone into a scientific expert, but enabling them to fulfil an enlightened role in making choices which affect their environment and to understand in broad terms the social implications of debates between experts. There is similarly a need to make everyone capable of making considered decisions as consumers. (pp. 11–12)

This extract highlights the 'democratic argument' for public understanding of science (Millar, 1996). Many commentators also stress the 'cultural argument', that some scientific ideas are a major cultural product, to which everyone ought to be introduced as part of their 'cultural inheritance'. Together these two arguments provide a rationale for content selection and design of science courses which aim to enhance students' scientific literacy.

2. A CENTRAL TENSION WITHIN SCHOOL SCIENCE EDUCATION

Whilst scientific literacy for all citizens is widely seen as an important goal for science education, the science curriculum also has to satisfy other demands. In particular, it has to provide the first stages of a training in science for those students who might later aspire to a career in science or who require more advanced knowledge of science. The tension between these two purposes of school science – scientific literacy for all and pre-professional training in science for some – is highlighted in the *Beyond 2000* report (Millar and Osborne, 1998). Many curricula, the report argues, attempt to achieve both purposes – with the pre-professional training emphasis invariably coming to have a dominant and distorting influence on the whole.

A major national project in England is currently exploring one way in which this tension might be addressed. The *Twenty First Century Science* course[1] is based on a curriculum model which separates the two purposes of science education: scientific literacy and pre-professional training of scientists. It divides the curriculum time allocation for science (20% is the norm for students aged 15–16 in

[1] *http://www.21stcenturyscience.org*

England) into two equal parts. All students take a Core Science course, designed explicitly to develop their scientific literacy. Alongside this they may choose to take an Additional Science course, which is offered with either a 'pure' or 'applied' emphasis. Both courses are of the same size as those followed by these students in other subjects leading to the General Certificate of Secondary Education (GCSE) qualification. Core-plus-Additional is designed so that it provides a sound basis for progressing to more advanced courses in the sciences at post-16 stage. Separating the courses allows us to develop each according to criteria which make them much more 'fit for purpose' than any single hybrid course, attempting to do both jobs, could be.

A partnership of the University of York Science Education Group and the Nuffield Curriculum Centre has worked with the Awarding Body OCR and the publisher Oxford University Press to develop detailed specifications for these three courses and programmes of study and teaching materials (textbooks, photocopy masters and e-learning materials) to support them. This has been piloted in 78 schools in England from September 2003. The first cohort of over 6000 students completed the two-year course in summer 2005 and obtained GCSE science awards.

3. TWENTY FIRST CENTURY SCIENCE: DESIGN PRINCIPLES

There is not space in this short article to describe fully the three courses which make up the Twenty First Century Science course (more information can be found in 21st Century Science Project team, 2003; Burden, 2005; OCR, 2005, Millar, forthcoming). In outline, the Core Science course interprets 'scientific literacy' in a manner similar to the authors of the European Commission White Paper cited earlier. All students (like all citizens) are 'consumers' of scientific information in their everyday lives – as they make personal choices (e.g. about health, diet, use of energy resources, etc.) or form views on issues affecting society (e.g. waste disposal, genetic modification of organisms, global climate change and CO_2 emissions). The aim of the course is to make them more intelligent and informed 'consumers', able to respond in an informed and appropriately critical manner to information they receive. This requires that they know some science and that they know some things *about* science as a form of knowledge and enquiry. The former we call *Science Explanations*, meaning a broad and usually qualitative grasp of the major ideas and explanations of science. The latter we call *Ideas about Science*; these include an awareness of issues concerning data and its interpretation, an ability to probe and evaluate evidence about trends and correlations, an understanding that theories and explanations are often conjectures that need to be checked against data and may change, an appreciation of the peer review process used within science to assure the quality of knowledge claims. We also include in the *Ideas about Science* category important ideas about the technological application of scientific knowledge, for example, an understanding of risk, and an introduction to ideas that are useful in deciding whether an application of science is acceptable or not (such as simple cost-benefit analysis).

The Additional (Pure) Science course is much more like 'traditional' science courses, but has a stronger emphasis on models and modelling in science, and makes use of the best research knowledge available on concept learning in key topic areas. The Additional (Applied) Science course has an emphasis on how science is used in the workplace, by technicians and others who use science routinely, rather than research scientists pursuing new scientific knowledge. It introduces students to standard procedures that are used in authentic situations, emphasising the need for careful measurement to collect valid and reliable data, for monitoring and testing purposes.

4. HOW WELL DOES IT WORK IN PRACTICE?

A fuller evaluation of several aspects of *Twenty First Century Science* is currently in progress, with reports due from the evaluators in September 2006, when the second cohort of students has completed the course. At present, evidence of the effects of teaching science in this way are based on more limited data, such as the accounts of teachers involved in the *Twenty First Century Science* pilot, collected through written questionnaires and interviews.

One question asked teachers to say if they thought the Core Science course was successful in improving students' scientific literacy, and to give their evidence for this. The following responses were typical:

Very successful with most students. Students are prepared to discuss a topic, question ideas and listen to others.

Clearly having an effect. Pupils discussing issues from experience, issues from news, from magazines, both in and out of lessons.

Students were amazed at first to be asked their opinions on topics. Now they are much more knowledgeable about current scientific issues and willing to express concerns, opinions.

We also asked teachers to compare their students' interest in science and engagement with school science with that of previous students, following more conventional science courses. The views of a sample of 40 teachers are summarised in the table below:

Much better	Better	Same	Worse	Much worse	No response
6	21	7	4	1	1

Many teachers attributed this to the motivating effect of clear links between the course topics and the science issues that students hear about in the news:

Most pupils are enthused about core and its [. . .] up-to-date approach, and take more interest . . .

More interest especially in science issues and will often comment on stories in media. Engagement real, as opposed to often tacit with traditional courses.

Some pointed out that this improvement applied also to students who do not normally achieve well in science:

Very pleased with the engagement of all abilities of pupils, with ethical issues such as . . . cloning, as these were issues that were familiar from press/TV.

Less able [students were] really engaged, in many cases [the topics] really gripped their imagination.

Of course, teachers also reported challenges and difficulties. Those mentioned most frequently were the language demand of some of the materials (for example, using extracts from newspapers and magazines), and how to stimulate and manage student discussion, particularly of open issues with no single right answer. These are challenges that are likely to arise in any scientific literacy course. We have tried to tackle the first of them by working collaboratively with pilot school teachers to revise printed materials to reduce the reading demand. And we are continuing to address the second by providing training for teachers in developing their teaching repertoire to include activities that involve more open discussion. It is encouraging to note that teachers' views of the course were even more positive at the end of the second year of the pilot than they had been at the end of the first.

CONCLUSIONS

The data summarised above come, of course, from teachers' self-reports of the impact of *Twenty First Century Science* on their practice and on their students' responses, provided to the project team who had developed the approach and the teaching materials. So we need to be a little wary in interpreting them. But they are nonetheless encouraging. The development of *Twenty First Century Science* has already showed that it is possible to design a school science course, for 15–16 year old students, with a distinctively different 'scientific literacy' flavour. Teachers clearly see the course as different in significant ways from previous courses. More important, many teachers welcome the ways in which it is different. They talk about these as 'how we should be teaching science' and 'how I've always wanted to teach'. So a 'scientific literacy' course can be designed, and it is attractive to many teachers. Also, from these teachers' reports, it is more popular with their students than previous science courses and stimulates greater interest and engagement with science. The key here seems to be the use of contexts drawn from the science that students hear about in out-of-school settings.

In the coming year, we hope to explore in more detail the impact of the *Twenty First Century Science* course on the attitudes towards science of students in the pilot schools, and on the numbers choosing to take science courses in the upper secondary school. We will report on these through the science education research journals when the data are available.

So, to answer the question in the title, school science *can* deliver scientific literacy to a much greater extent than it does at present. The key is to recognise and deal effectively with the two purposes of school science: scientific literacy for all, and the pre-professional training of scientists. If we do not, we risk continuing to offer a course that turns the majority off science, and provides a poor foundation for advanced study for those who choose to follow that career path. By separating the two functions of school science, we can do both better. The importance of *Twenty First Century Science* is that is this claim is no longer simply a professional judgment but one that is increasingly supported by evidence.

ACKNOWLEDGEMENTS

Substantial grants from the Nuffield Foundation, the Wellcome Trust and the Salters Institute made possible the development of the *Twenty First Century Science* course and teaching materials.

REFERENCES

21st Century Science Project team (2003) 21st Century Science – a new flexible model for GCSE science. School Science Review 85(310):27–34

Burden J (2005) Twenty first century science. Education in Science 213:10–12

European Commission (1995) Teaching and learning. Towards the learning society. White paper on education and training (COM(95) 590). Brussels: European Commission

Goodrum D, Hackling M, Rennie L (2001) The status and quality of teaching and learning of science in Australian schools. Canberra: Department of Education, Training and Youth Affairs

Harlen W (2001) The assessment of scientific literacy in the OECD/PISA Project. Studies in Science Education 36:79–104

Layton D, Jenkins E, Donnelly J (1994) Scientific and technological literacy. Meanings and rationales. An annotated bibliography. Leeds: Centre for Studies in Science and Mathematics Education, University of Leeds, in association with UNESCO

Layton D (1975) Science for the people. London: George Allen & Unwin

Millar R (1996) Towards a science curriculum for public understanding. School Science Review 77(280):7–18

Millar R (forthcoming). Twenty first century science: Insights from the design and implementation of a scientific literacy approach in school science. International Journal of Science Education (in press)

Millar R, Osborne J (1998) Beyond 2000. Science education for the future. London: Nuffield Foundation

National Research Council (1996) National science education standards. Washington, DC: National Academy Press

OCR (Oxford Cambridge and RSA Examinations) (2005) Twenty first century science suite. OCR GCSE Science A (J630). Specification. Cambridge: OCR. <available online at: *www.ocr.org.uk*>

OECD (Organisation for Economic Cooperation and Development) (1999) Measuring student knowledge and skills. Paris: OECD. <available online at *www.pisa.oecd.org*>

UNESCO (1990) World Declaration on Education for All. <available online at: *http://www.unesco.org/education/efa/ed_for_all/background/jomtien_declaration.shtml*>

CHAPTER 26

PERCEPTIONS AND IMAGES OF SCIENCE AND SCIENCE EDUCATION

Some simple results from ROSE – a cross-cultural study

SVEIN SJØBERG[1] AND CAMILLA SCHREINER[2]

[1]*Department of Teacher Education and School Development, University of Oslo, Norway, PO Box 1099 Blindern, 0317 Oslo, Norway, Tel +47-22-854155, Email svein.sjoberg@ils.uio.no*
[2]*Department of Teacher Education and School Development, University of Oslo, Norway, PO Box 1099 Blindern, 0317 Oslo, Norway, Tel +47-22-844397, Email camilla.schreiner@ils.uio.no*

Abstract: Modern societies become more and more dominated and driven by science and technology (S&T). Such societies need a highly skilled workforce for a competitive global knowledge-based market. Furthermore, democratic societies need broadly educated citizens who can understand current S&T-based challenges and who can take part in democratic decisions. In the EU and many other countries there is widespread concern about a fall in the recruitment to studies and careers in S&T. In order to face these challenges, we need to know how young learners perceive S&T in society; we need to know their attitudes, sentiments and priorities. This chapter gives a brief glimpse into this situation by presenting some data from ROSE, an international comparative research project. The data indicate that young students actually have great respect for and many positive attitudes towards S&T. This does, however, not imply that the want to choose S&T as studies and careers. The data also shows great differences between girls and boys in attitudes, values and priorities. Any attempt to redress the situation will need to take such knowledge into account in order to be successful

Keywords: Science education, attitudes to science, perceptions of science, interest in science, recruitment to science

1. INTRODUCTION

In many highly developed countries one can notice a declining recruitment of students of science and technology (S&T) studies. "Europe needs more scientists!" is the title of the report from the High Level Group addressing the condition of

M. Claessens (ed.), Communicating European Research 2005, 149–156.

science and technology in the EU, with special attention to the number of people entering S&T educations and careers (European Commission, 2004). The title of the report reveals the point: The falling number of students of most S&T education is seen as a large problem in most European countries. The same tendencies are noted in the US (NSB, 2004) and in most other OECD-countries.[1] This chapter presents data and results provided by an international comparative project, ROSE – the Relevance of Science Education.

There are large and interesting differences between the countries, both with respect to the number of students, which subjects have the lowest recruitment, how large the recruitment problem is perceived to be, etc. The "hard" S&T-subjects such as technology, engineering, physics and to some extent chemistry are particularly affected. In addition, the gender differences vary from one country to another. In most countries, although with large variations, the boys outnumber the girls in physics and engineering studies, while the gender balance swings towards the girls in studies like medicine, veterinary medicine, environmental science and biology. We consider the lack of perceived relevance of the S&T curriculum to be the main reason for young peoples' low interest in the school subject and their lack of motivation for pursuing the subject in their higher education. With *meaning* and *relevance* as key words, we have developed the ROSE project – The Relevance of Science Education.

2. ROSE IN BRIEF

ROSE, the Relevance of Science Education, is an international comparative project meant to shed light on affective factors of importance to the learning of science and technology. The target population is students towards the end of secondary school (age 15). The research instrument is a questionnaire mostly consisting of closed questions with four-point Likert scales. The advantage of using this scale is that it is commonly treated as a quasi-interval scale. This implies that we can use standard statistical methods like means, correlations etc. The rationale behind the project, including the questionnaire development, theoretical background, procedures for data collection, etc. is described in Schreiner and Sjøberg (2004), and is available in print from the authors or from the project website.[2]

The key feature of ROSE is to gather and analyse information from the students about several factors that have a bearing on how young people relate to S&T at school and in society and their motivation for learning S&T. By means of international discussions, workshops and pilot studies with many research partners, ROSE has d developed an instrument that aims to map out the attitudes of

---

[1] This chapter draws on material from Schreiner, 2006; Schreiner & Sjøberg, 2004, 2006

[2] *www.ils.uio.no/forskning/rose/*

15 year old students towards S&T in education and in society. Research groups from about 40 countries give their input to ROSE. The aim is to stimulate research cooperation and networking across cultural barriers and to promote an informed discussion on the means to make science education more relevant and meaningful for students in ways that respect gender differences and cultural diversity. We also hope to shed light on the manner in which we can stimulate the students' interest in choosing S&T-related studies and careers – and to stimulate their life-long interest in, and respect for, S&T as part of our common culture. About 10 PhD students are preparing their thesis on the basis of ROSE data, and the first PhD-thesis was completed in Norway in November 2005 (Schreiner, 2006).

Here, we can only report on some simple results from an analysis of the ROSE material, confining ourselves to show mean scores for girls and boys from a number of countries in the ROSE sample. The countries are sorted partly geographically, with neighbouring countries together; and partly by level of development, with more modernised countries at the bottom. We use the Human Development Index (HDI) as a measure for the level of development of a country (UNDP, 2005). We only include countries where the sample meets certain quality criteria that guarantee that the sample is representative.

The Likert scales have four response categories, and the response categories are similar from one question to another, going from 'small' (coded 1) to 'large' (coded 4). This means that average scores of 2.5 imply that the students in the country in average are neutral to the statement; that they neither agree nor disagree.

3. MOST OF YOUTH APPRECIATES S&T

A possible explanation for young people's lack of interest in studying S&T could be that they hold a negative view of the role that S&T play in society, and that they blame S&T for the unintended catastrophes and risks (e.g. the Chernobyl disaster in 1986, the Bovine Spongiform Encephalopathy or "mad cow disease", the depletion of the ozone layer, global warming and overpopulation) following in the wake of the technological development. Contrary to such expectations, the ROSE results indicate that young people in most parts of the world express a positive view of S&T. Average scores for girls and boys in nearly all countries show strong agreement with statements like these:

- Science and technology will find cures to diseases such as HIV/AIDS, cancer, etc.
- Science and technology are important for society.
- Thanks to science and technology, there will be greater opportunities for future generations.
- New technologies will make work more interesting.
- The benefits of science are greater than the harmful effects it could have, and
- Science and technology make our lives healthier, easier and more comfortable.

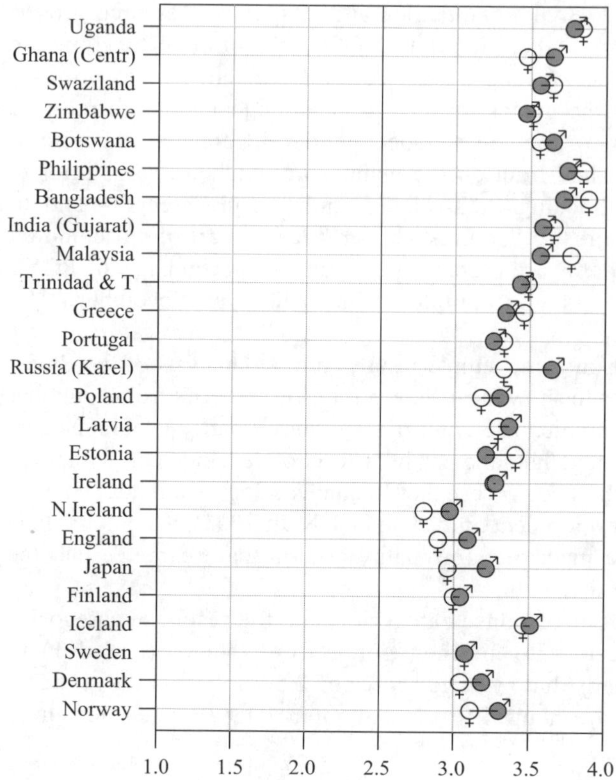

Figure 1. "Science and technology are important for society." Average scores for boys (filled symbols) and girls (open symbols). (1 = Disagree, 4 – Agree) Brackets against the country's name, indicates that the country's data was obtained via a sample obtained from a region of that country: i.e. Ghana has data from the Central region, India has data from Gujarat, and in Russia the data are collected from Karelia. "Trinidad & T" denotes Trinidad and Tobago

Figure 1 illustrates this with one example. The diagram shows responses to "Science and technology are important for society". In average, girls and boys in all countries agree that S&T are important for society, and the gender differences are negligible.

4. . . . BUT DO THEY WANT A JOB IN S&T?

Some questions address this issue. Figure 2 shows that there are large cross-national differences when it comes to students' agreement with the statement "I would like to become a scientist." The mean scores in the developed countries are extremely low, and the girls are even more negative than the boys. Japan has particularly large gender differences.

Responses to the item "I would like to get a job in technology" are illustrated in Figure 3. Also in this diagram, we note pronounced differences between countries

Figure 2. "I would like to become a scientist." See caption 1 for explanation

and between girls and boys in each country. While boys in more developed countries give average scores close to the neutral value, most girls in these countries do *not* want to work with technology. In developing countries, both girls and boys agree with the statement. Also in these countries, there are some gender differences, but they are by no means as large as in the developed countries.

5. YOUTH'S ORIENTATION TOWARDS S&T – LINKED TO THE LEVEL OF DEVELOPMENT IN A COUNTRY

Young people's values, views and ways of understanding themselves, their surroundings and the world are products of the culture in which they are growing up. Our data show a strong co-relationship between the Human Development Index (HDI) for a country and the responses in the ROSE questionnaire. One example is given in Figure 4, which shows the national average score across all the 108 items in the questions "What I want to learn about") on one axis and the HDI on the other. The Pearson product moment correlation coefficient is $-.85$ ($p < .01$), which indicates

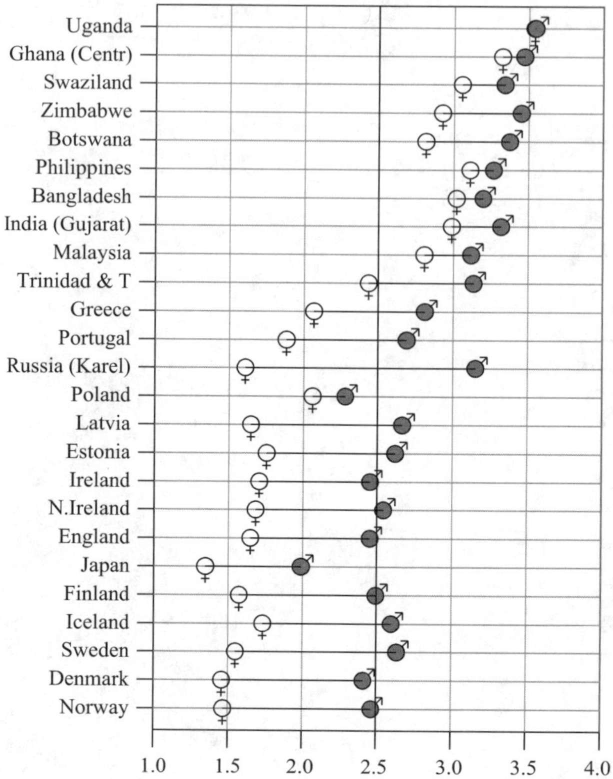

Figure 3. I would like to get a job in technology. See caption 1 for explanation

a very strong inverse relationship: The higher the level of development in a country, the lower the interest the students express in learning about S&T-related topics.

Figure 4 shows that the mean for some of the countries in the extreme low end of the HDI scale lies one unit (in a scale with a span of three units) above many countries in the extreme high end of the HDI scale. The correlation coefficients for some of ROSE questions are given in Table 1. The table indicates the general pattern which emerges from the analysis of the ROSE material: The more developed a country is, the less positive young people are towards the role of S&T in society and their own willingness to get involved in S&T careers and studies.

We see from our data that whether young people wish to opt for a career in S&T is closely related to the country's level of development. An important challenge in poor countries is, of course, related to the improvement of material conditions, economic growth and to a better health and the welfare system. Further material development of the society is naturally a main political and public issue, and in this respect, S&T are seen as fundamental driving forces. One may assume that in such

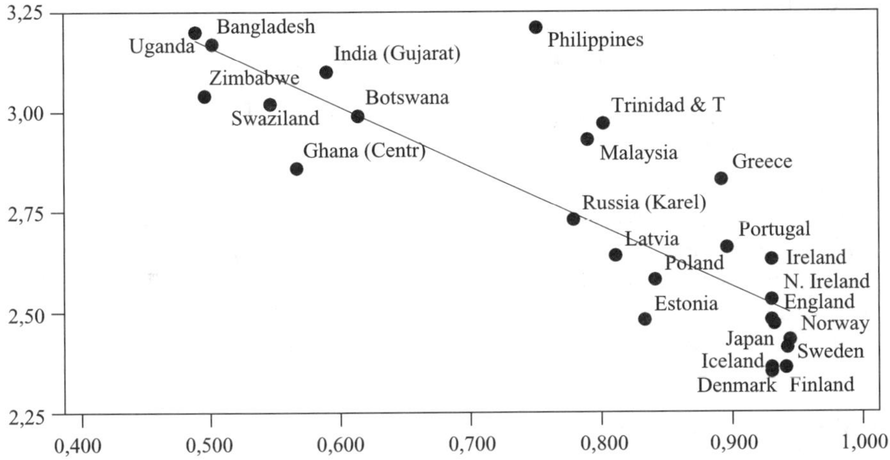

Figure 4. Scatter-plot with regression line: HDI values (horizontal axis) and the national average score (on the 4-point scale) across all 108 items in the question "What I want to learn about" (vertical axis) for all countries

societies, a job in S&T is perceived as important for society and thereby as meaningful for the individual.

When today's modern societies were in the era of early industrialisation, the focus was directed towards *progress*, *growth* and *building the country*. Consequently, exactly this – to build the country – was perceived as important for the society and meaningful for the individual. It may be that we now have passed the era in which the work of physicists, technicians and engineers are seen as crucial for people's life and well-being in the more developed countries.

Other studies also indicate that in poorer countries, young people have a rather heroic image of scientists as persons, while this is not the case in highly developed Western societies (Sjøberg, 2002). In modern societies, neither scientists nor engineers are heroes or attractive role models for the young generation. Obviously, the level of development influences people's hopes for the expected benefits of developments in S&T. The recent Eurobarometer survey (European Commission, 2005) also shows that the belief in the benefits of S&T is much stronger in the less developed EU countries than in the wealthier and more developed countries.

Table 1. Pearson product moment correlation of country mean responses with the HDI of the countries. All correlations are significant ($p < .01$)

– I would like to become a scientist	0.94
– I would like to get a job in technology	0.91
– Science and technology are important for society	0.78
– The benefits of science are greater than its potentially harmful effects	0.73

CONCLUSIONS

The above results raise important questions about possible strategies to encourage young people in highly developed countries to find S&T studies and careers attractive. Our own contention (which we can substantiate with data as well as theory) is that young people need to see the personal relevance of S&T in their own life, their values and their identity. If they cannot find this relevance, they would rather opt for one of the many other possible futures and careers that life in a rich and developed society has to offer. We, as science communicators, have to listen to the values, hopes and aspirations of the young if we want them to study and work in S&T. The ROSE project has a wealth of empirical evidence on which we can start to get engaged in such necessary critical appraisals of our approaches to S&T curricula and teaching practices.

REFERENCES

European Commission (2004) Europe needs more scientists! Brussels: European Commission, Directorate-General for research, high level group on human resources for science and technology in Europe. Available from: http://europa.eu.int/comm/research/conferences/2004/sciprof/pdf/final_en.pdf (accessed 2005-10-24)

European Commission (2005) Europeans, science and technology. Special Eurobarometer224, Brussels: European Commission (Available from http://europa.eu.int/comm/public_opinion/index_en.htm)

NSB (2004) Science and engineering indicators 2004 (NSB 04-01). Arlington, VA: National Science Board, National Science Foundation. Available from: www.nsf.gov/nsb/documents/reports.htm (accessed 2005-10-24)

Schreiner C (2006) Exploring a ROSE-garden. Norwegian youth's orientations towards science – seen as signs of late modern identities. Based on ROSE (The Relevance of Science Education), a comparative study of 15 year-old students' perceptions of science and science education. Doctoral thesis, University of Oslo, Faculty of Education, Department of Teacher Education and School Development, Oslo

Schreiner C, Sjøberg S (2004) Sowing the seeds of ROSE. Background, Rationale, Questionnaire Development and Data Collection for ROSE (The Relevance of Science Education) – a comparative study of students' views of science and science education (Acta Didactica 4/2004). Oslo: Dept. of Teacher Education and School Development, University of Oslo. Available from: www.ils.uio.no/forskning/publikasjoner/actadidactica/index.html (accessed 2005-08-06)

Schreiner C, Sjøberg S (2006) Science education and youth's identity construction – two incompatible projects? In: Corrigan D, Dillon J, Gunstone R (eds) The Re-emergence of values in the science curriculum: Sense publications (in progress)

Sjøberg S (2002) Science for the children? Report from the science and scientists-project — Acta Didactica. -(1/2002) Dept. of Teacher Education and School Development, University of Oslo

UNDP (2005) Human development report 2005. New York: United Nations Development Programme. Available from: http://hdr.undp.org/ (accessed 2005-09-10)

CHAPTER 27

REPRESENTING SCIENCE THROUGH MULTIPLE-CHANNEL DIGITAL TELEVISION

Opportunities for dialogue, engagement and deliberation?

RICHARD HOLLIMAN

Faculty of Science, The Open University, Walton Hall, Milton Keynes, MK7 6AA, United Kingdom, Tel +44-1908-654646, Email R.M.Holliman@open.ac.uk

Abstract: Television has long been considered an important medium for communicating science, not least because television programmes have the potential to reach large audiences, disseminating information about the work of a range of scientists working in a variety of disciplines. This paper briefly considers recent developments in two important areas – technological innovations (e.g. the shift from analogue to digital television) and deregulation (e.g. the increasing take-up of multi-channel television) – critically evaluating the potential for these developments to provide new opportunities for dialogue, engagement and deliberation

Keywords: science communication; digital television; dialogue, engagement and deliberation

1. INTRODUCTION

Television is an important source of information about newly published scientific research for citizens once they have completed formal school-based education.[1] Portrayals of newly published scientific research are likely to be partial, however, as studies have noted how a range of actors, including media professionals, scientists and non-governmental agencies, mediate information within particular cultural and institutional circumstances, and following certain specialised production values

[1] House of Lords Select Committee on Science and Technology, 2000.

M. Claessens (ed.), Communicating European Research 2005, 157–162.

(e.g. Holliman, 2004). Understanding how audiences interpret and contextualise these representations is a complex task, requiring methods that allow viewers to articulate their views within a familiar social context and using their own vocabulary (Holliman, 2005). It can be argued, therefore, that television portrayals currently have the *potential* to engage a wide range of citizens with partial, mediated representations of complex science-based issues, but that media influence of these portrayals cannot be easily predicted.

This comes at a time when the relationship between science and society is being reconsidered in the UK and other EU countries (for an overview of these issues, see Miller, 2001). Widespread criticism of the "deficit model" throughout the late-1980s and 1990s (see Irwin and Michael, 2003 for discussion) was accompanied by calls for greater openness and transparency about science-based issues, alongside the introduction of opportunities for dialogue, engagement and deliberation. More recently, calls have been made to move these deliberations "upstream", that is, to involve members of the public in decision-making processes about complex science-based issues as they emerge (Willis and Wilsdon, 2004), then continuing these deliberations throughout the development of a given issue.

On a similar timescale to these ongoing critiques, the introduction of new technologies, such as digital television, which offers interactivity in real-time, and developments in the broadcast media marketplace, such as multi-channel TV, which allows greater audience choice, have converged to provide new ways of consuming science and other forms of programming. This paper briefly investigates whether multi-channel digital television has the potential to provide effective opportunities for real-time audience engagement and deliberation about complex science-based issues.

2. THE ROLE OF NEW TECHNOLOGIES AND DEREGULATION: FROM VIEWERS TO USERS?

Recent figures from the media regulator Ofcom state that nearly two thirds of UK households have access to digital TV (Deans, 2005). It is within this context that the decision has been made to "switch off" UK analogue TV signals region-by-region between 2008 and 2012. It is not surprising, therefore, that promotional campaigns to encourage remaining analogue viewers to switch to digital TV, which can be accessed through satellite, cable and freeview technologies, continue to be broadcast, whilst more and more programmes take advantage of the opportunities to engage interactively with viewers. You may have even pressed the red button on your digital remote control to access additional services. What then are the advantages for communicating science on digital television?

Educational technologists have evaluated different media as a way of gauging their pedagogic potential to facilitate science (and other) learning (Laurillard, 2004). In this sense, analogue TV can be seen as a linear medium, where producers use a number of tools to structure programmes in ways they hope viewers will understand and enjoy. Digital TV does this too, but also has the potential to offer interactivity in real-time in the sense that viewers can actively request additional information, potentially even voting in real-time to influence the direction of programmes, or deliberate about a given issue. In this way *viewers* can become *users*. However, in doing so, it is important to remember that viewers need to be motivated to engage with these new opportunities for interaction, which relates, in part, to ideas about consumer choice.

Deregulation, increasing the influence of free market economics in the broadcast media marketplace, has played an important role in shaping consumer choice. For example, the introduction of multiple-channel TV in the early 1990s has influenced how viewers consume programmes, reducing audience share between 1991 and 2004 for BBC1 (by 9.3%) and ITV1 (by 19.2%), while audience share for non-terrestrial channels grew by 22.2% (BARB, 2005). The overall result of deregulation, therefore, has been an increase in the overall number of channels, resulting in opportunities to narrowcast to particular audiences about specific topics, including science-based issues, and a reduction in audience share for the two main free-to-air channels (BBC1 and ITV1).[2] Key to all this is the ongoing tension, particularly for the main terrestrial TV companies, between consumer choice and fulfilling public sector broadcasting (PSB) requirements. Over time, however, the lessons from radio suggest that PSB remits may diminish as deregulation takes an ever greater hold. Taken together, this means that science will both have to compete to maintain its current position in broadcast schedules and in viewers' consumption patterns.

3. PROMOTING SCIENTIFIC CITIZENSHIP ON TV: THE CASE OF ORGAN DONATION

The increasing switch to digital TV offers opportunities for real-time audience interactivity, providing viewers (citizens) with chances to engage with and deliberate about complex science-based issues. A recent example of this formed part of the BBC's recent (summer 2005) week-long DoNation season, which portrayed some of the scientific, social and ethical aspects of the complex biomedical-based issue of organ donation. A key reason for this activity was the lack of suitable organ donors in the UK, which has the lowest number of potential donors in Western Europe (Lyall, 2005). More than 6 000 patients are currently (2005) waiting for a transplant in the UK, while only 2242 transplant operations were completed in 2004 (ibid.); 452 died waiting.[3]

A key aspect of the DoNation season was the joint venture between two popular TV medical dramas – *Casualty* and *Holby City*, resulting in *Casualty@Holby – Interactive*. This docudrama, entitled 'Something we can do', is most easily described in four parts, the first being a factual introduction to organ donation, the aims of

---

[2] Digital television has the potential to develop consumer choice even further with the introduction of "on demand" viewing and programme downloads.

[3] *Casualty@Holby – Interactive*, BBC1, 27/08/05.

the DoNation season and the programme itself by the high-profile scientist and peer, Professor Robert Winston.

The second part of the programme took the form of a drama set in the Holby City Hospital, involving three converging storylines. The first portrayed Matt, a young man estranged from his parents for 11 years, who was involved in a serious car accident on the way to a family reunion. Viewers saw how the casualty departments' medical professionals struggled in vain to save Matt's life. When this character was subsequently declared brain stem dead his estranged parents had to come to terms with his death, and meet his pregnant wife for the first time. All three characters were then asked to consider the prospect of organ donation – Matt carried an organ donor card – as they were introduced to the hospital's donor coordinator. The deliberations between these three characters, medical professionals, family members and friends, and the donor transplant coordinator were a key element in this developing narrative, encompassing biomedical, social and ethical issues surrounding organ donation. In the end, Matt's wife, as his next of kin, formally agreed for his organs to be donated.

Running alongside Matt's story, viewers were introduced to Lucy, a young cystic fibrosis patient, who was waiting for a heart and lung transplant operation. She was also rushed to hospital after having collapsed in a pub. The casualty staff managed to stabilise Lucy, who then described her frustration at not being able to lead a "normal" life, both to medical professionals and her parents. The latter were strong advocates that Lucy should receive Matt's transplanted organs.

The third storyline featured Tony, who had a severe heart condition requiring a heart transplant. Approximately 20 years older than Lucy, he also collapsed and was rushed to hospital, where he had to be revived after his heart had stopped. Having lost his first wife and child in a car accident some years previously, he got engaged in the hospital when his girlfriend proposed. His new fiancé was a strong advocate that Tony should receive Matt's transplanted organs.

The third part of the programme offered viewers the chance to vote – by phone – on which of the two endings of the episode they wished to view, i.e. who received Matt's organs. At this point the programme became a documentary, describing medical practitioner guidelines for matching organ donors with recipient patients: medical urgency, matching body size, blood group and tissue type, the timing and distance for retrieval and delivery of organs for transplant, and life expectancy of the recipient patient were described in turn. A series of vox pops then followed, expressing a range of opinions about organ donation. Four common concerns were then described: "tempting fate"; the idea that doctors might not try as hard to save a patient's life if they had signed the organ donor register; how can relatives and friends be confident that a potential organ donor is really dead?; what will an organ donor look like after the retrieval operation? Finally, the parents of a nine-year-old child whose organs were donated were interviewed, reflecting on how they fulfilled their son's wishes (he carried a donor card).

The fourth part of the programme showed the results of the phone-in. Of the 7.32 million viewers who watched the programme, 98,800 voted. Of these, 65% voted for Lucy and 35% for Tony. Viewers also watched the ending chosen by the majority, seeing Lucy complete a university degree in apparently good health and Tony getting married in a wheelchair.

Finally, at the end of the programme, digital viewers were informed that they could also access further information about organ donation and becoming a donor by pressing the red button. Furthermore all viewers, whether analogue or digital, where provided with BBC web site addresses for further information about organ donation and signing the organ donor register, or to view the alternative ending where Tony was the recipient of Matt's organs. Viewers were also provided with a free-phone telephone number in case they had been affected by issues discussed in the programme.

Overall, a key aim of the programme was to highlight the difficult decisions that medical professionals have to make due to the lack of suitable organ donors, promoting engagement with these complex issues by giving viewers the chance to make such a decision, albeit following a contrived scenario. Viewers were then offered the chance to become part of the solution by volunteering to become organ donors themselves through the associated online DoNation campaign; an example of real-time deliberation potentially influencing citizen engagement with this issue.

How then can we judge the success of such a programme? Viewing figures provide only a partial measure of success of programmes of this nature. It could be argued, for example, that 7.32m viewers watched, and therefore engaged at some level with these issues; an impressive figure in keeping with audience figures for these programmes. In this sense, it can be inferred that the programme engaged viewers who were already likely to watch hospital-based dramas that address complex biomedical issues in context. But how do we make sense of the 98,800 viewers who chose to vote for a particular ending – can this be considered a "successful" deliberation? If so, should it inform public policy on an issue of this kind? Surely a more effective measure in this instance would be to systematically investigate the views of those who watched the programme, sampling those who subsequently did and did not sign up to the organ donor register, and/or inform their next of kin of their wishes.

CONCLUSION

This short paper has described some of the potential opportunities for those wishing to communicate science in an age of multi-channel digital television. Overall, developments in technology and increased deregulation suggest that real-time audience engagement and deliberation *could* become a routine aspect of watching science on television. Does all this sound too good to be true? Certainly new technologies can extend the opportunities for dialogue and engagement, but this is not without its challenges. For example, the move towards greater access to on-demand multi-channel TV is likely to provide opportunities for more dynamic, interactive science-based communication: potentially through convergence with new technologies, such as the Internet, MP3 players and mobile phones, but with smaller audiences. As a result, viewers will have ever greater choice. Motivating citizens to consume science will become a crucial factor if this programming is not to be relegated to a niche market of enthusiasts on channels with a strong PSB remit and/or a well-established interest in science-based issues.

REFERENCES

Broadcaster's Audience Research Board LTD. (BARB) (2005) Annual percentage shares of viewing (1981–2004); *http://www.barb.co.uk/index1.cfm/flag=home*;

Deans J (2005, December 9) Digital TV enjoys growth spurt, Guardian Unlimited; *http://media.guardian.co.uk/broadcast/story/0,1663791,00.html*;

Holliman R (2005) Reception analyses of science news: Evaluating focus groups as a method. Sociologia e Ricerca Sociale **76–77**:254–264

Holliman R (2004) Media coverage of cloning: A study of media content, production and reception. Public Understanding of Science **13**:107–130

House of lords select committee on science and technology (2000) Science and society, Third Report. HMSO, London, 23 February

Irwin A, Michael M (2003) Science, social theory and public knowledge. Open University Press, Buckingham

Laurillard D (2004) Rethinking the teaching of science. In: Holliman R, Scanlon E (eds) Mediating science learning through information and communications technology. RoutledgeFalmer, London, pp:27–50

Lyall J (2005) Press red button, donate kidney, StudentBMJ **13**:393

Miller S (2001) Public understanding of science at the crossroads. Public Understanding of Science **10**:115–120

Willis R, Wilsdon J (2004) See through science: why public engagement needs to move upstream, DEMOS, London

CHAPTER 28

HOW TO GET SCIENCE IN THE NEWS

A true story

JAN-OLOV JOHANSSON

Senior producer Science Department Swedish Radio, Fältvägen 17b, SE-756 46 Uppsala, Sverige, Tel +46-70-8531122, Email jan-olov.johansson@bahnhof.se

Abstract: Swedish Radio is nowadays broadcasting three daily science news shows during prime time, but this has not always been the case. In fact, ten years ago very few people believed that there would be any public interest in science at all. Even fewer would have guessed that *Science News* would be among the highest public ratings and lead the list of downloaded pod radio. This chapter tells the story about how *Science News* was established. One of the bigger obstacles to get science in the news was not the lack of curiosity among the audience but rather the lack of interest of colleagues in the media. Traditionally, people working in media have a non-scientific education and it is therefore possible to see this as a confrontation between the two cultures, first described by C. P. Snow. As a journalist dealing with science you have to work on two fronts, the first being to tell the public how crucial development in science is for their daily life and the second to convince colleagues within the media that science is an essential part of our society today. Europe, for several reasons, needs to cultivate professional science journalism and build a network among people working in the field. This work is essential as the research of today will be on the front page tomorrow

Keywords: Science news, Science in radio, Cultures, Public interest in science

Swedish Radio (SR) today has a unique position in the radio world when it comes to reporting about science. With three daily science news broadcasts during peak hour, the *Science News* programme of SR reaches over 1 million listeners. The science news is also highly rated among the public and is one of the most downloaded programmes on the pod radio. Over the last ten years, Swedish Radio has doubled its science broadcasts. This is the story about how that happened.

As a starting point, here is some information about radio in Sweden. Private radio started late in the country and still consists merely of stations playing

M. Claessens (ed.), Communicating European Research 2005, 163–167.

different categories of music. Swedish Radio (SR) is a public service institution, similar to BBC or NPR and dominates the radio in the country. Most Swedes (86%) listen to SR every day, especially in the mornings and around noon. SR broadcasts four main channels or programmes, accordingly called P1, P2, P3 and P4. The last channel, P4, has around 20 offices spread over the country to cover local events. Science Radio traditionally has been broadcasted in P1, together with programmes about politics, economy, culture and religion. Contrary to popular belief, radio is not a dying relic – instead, it finds a new audience via cell phones, pod cast and web cast.

1. TEN YEARS OF DEVELOPMENT

When I came to the Swedish Radio in November 1992, the science department was a sleepy academic branch producing a few shows a week and typically a feature programme about a timeless scientific subject.

I was given a limited time to get the department to develop – otherwise it would be closed down. I started out with two main ideas.

Firstly, people are very interested in science, technology and medicine – if you are able to tell them about it in a way that relates to their everyday life. You should always start from the public interest and tell the story from their perspective, not from an academic point of view. We are journalists working with science – not scientists!

Secondly, no modern society exists independently of science and technology. They are essential parts of our culture and should be respected as such. When there is a public discussion about economy, politics or society, you should always include science.

To prove those points, we started a weekly one-hour show that focused on how science is relevant to society, but it also told fantastic stories about outer space and stem cells. This worked out fine but we were still outside the public attention span. To really be one of the subjects people would discuss at breakfast or during coffee break at work, you had to take it one step further. You had to be in the news. News is media's "raison d'être". It is also a special form of craft of the news journalists, so the easiest way to get science into these broadcasts would, of course, be to feed those professionals new findings and exciting discoveries. Or so we thought. But to our surprise, very few of the items we found exciting made it to the listeners. In the end, news involving science, technology and medicine were often deemed to be "not newsworthy". One common statement for not running one of our stories was that "the public are not interested". Going about it scientifically, we started asking for proof of this "lack of interest". In the search for answers we also met colleagues from Germany, France, Great Britain and USA. We even organized international seminars about science and media in Uppsala. They all told the same story. The biggest obstacle in bringing science to the public is not the lack of interest among the public, but rather the lack of education among the gatekeepers of the airwaves!

This is so because the typical gatekeeper has no scientific background or interest. He or she was keener on politics, economy, art and literature. C. P. Snow's old remarks about the two cultures felt very relevant. So – how could we change that attitude? It turned out that the seminars worked well also from this aspect. We invited gatekeepers to give them a glimpse of what an unexploited goldfield they had overlooked. Some of them saw the opportunities. However, the change was far too slow. After the first six years, we drew the conclusion that if you want to get science into the news you have to do it yourself. But to get a slot in media you have to fight for it. First you have to prove that the public really wants what you are offering. Usually, this is done by public surveys. When looking at surveys, you will soon find that they are often pre-designed to give the answers the designer would like to obtain. At the radio we did not even ask the public about their interest in our subjects. In fact, just getting the word "science" into the questionnaire of 1993 was our first step towards success.

2. THE TURNING POINT

The result of the first minor surveys, carried out by SR, followed by others that all told the same story, showed that people who listened to channel 1 (P1) were quite fond of science and wanted to hear more about it. These findings were not understood by representatives of other departments, who made the observation that "there must be something wrong with the survey". But it is hard to argue with facts and after much lobbying over a period of six years, we finally got our chance in 1998 in the news. Only three minutes each morning, less than the Christian morning service, but still. Now there was the challenge of recruiting a group of dedicated professionals who are interested in science. And you needed a critical mass of these talented persons to create an active discussion.

We were lucky to find such people, and if the media world was sceptical, our listeners were not. Many people contacted us with questions, support and comments. So, when yet another survey was made after a few years, it did not surprise us that we actually won the competition. The people who listened to channel P1 wanted more science (Figure 1). In fact, they wanted science more then all other areas presented to them in the morning. But more than that, they also rated that the science news was the best quality in the morning broadcasts!

These results have since been recurring in other media and colleagues have slowly been convinced. The local stations (P4) started to relay our news to their listeners. That means that today, *Science News* reaches over 1 million people in Sweden. That is not so bad considering Sweden has a population of 9 million and that no-one believed in the idea just 8 years ago. When pod cast started, science was at first not included. As soon as it was, *Science News* went to the top of the ten most downloaded programmes list and stayed there. There can be no doubt that the public in Sweden are interested in science, if it is presented to them in the right way.

At the same time, we started many new feature programmes about science. This kind of longer programme becomes even more important when you do news. We all need to get some general knowledge on the subject and a bigger perspective, so as

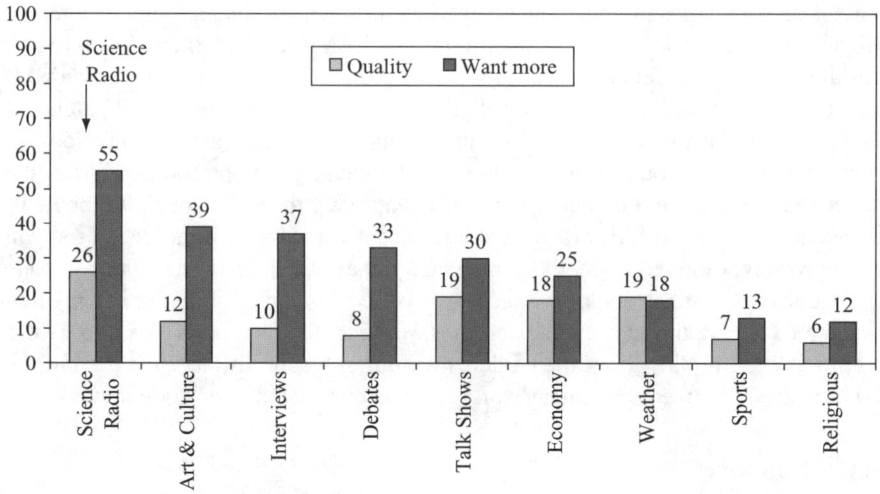

Figure 1. Interest of Swedish Radio listeners in various broadcasts

to not be blinded by all the news flashes. When I moved on, exactly ten years after starting as the editor in charge, the science department had more than doubled its scheduled airtime and reached over 1 million listeners via its news. Science had become a part of the public agenda.

3. WHAT IS NEWS ANYWAY?

There are many definitions of what is newsworthy. The perhaps oldest saying is that "if a dog bites a man that's not news, but if a man bites a dog, that's news". Another way of putting it might be that something unexpected has happened. However, that argument is built on the assumption that we all know what is "normal" – that is that dogs bite men, that we all have a pre-understanding of the situation. This is not always the case with science when you now and then get news that is really new in the sense that no-one ever heard of it before! People in general have no clue what a black hole or a stem cell is, nor does the typical editor in media. You therefore face the double challenge of educating both the listeners and your boss. As said above, it is hard work and you can not help thinking of C. P. Snow's speech about the two cultures. If science today is the work of persons who have an inherent scientific education, it is not true in the media. It is a paradox that means that science is not covered according to its importance for society in the general media.

4. TO TELL A STORY

All humans are born curious. We all love a good story. If you are able to tell the fascinating tale about how the borders of our knowledge expand almost every day it is not really a challenge to get people to listen. But how can you tell what the public wants? In commercial press the answer is relatively easy; if people buy your product, it is right. The problem with scientific magazines has traditionally been to find advertisers to help pay the production and distribution costs, a problem which is often solved with ads for academic positions. In spite of that, there have been several great business successes in popular science magazines during the last few years.

When you work as we do, in a big editorial office and compete for space or airtime with many other departments, it is a different story. The obstacle here is rather the personal interest of the editor than of the public. However, our experience has taught us a few things.

If you are serious about reporting, scrutinizing and commenting modern science you need a space of your own, where you can be the judge of what is newsworthy or not, very – much as is the case for art or culture pages, such as those in the most serious newspapers. To run this space professionally you also need a critical mass of journalists that know the area and know the art of telling a story. You need a group of critically trained journalists interested in science.

Scepticism and a critical eye are especially important in science journalism. You have to cultivate this constantly. In order to get funding for their research, or just to gain status, scientists have been known to promise results they can not deliver, or to simply cheat. Stem cell research is an example of an area where science as well as media sometimes have been promising cures too many times and too fast. The danger of overselling is bigger if you as a journalist have to fight an uninterested and badly educated gatekeeper on your way. Journalists covering science are now and then tempted to oversell their stories just to pass the aforesaid gatekeeper.

5. FINAL WORDS

To conclude here are some remarks drawn from our experience.

- Europe needs a vital public discussion about the essential role of science, technology and medicine plays in society.
- Media has got to play an active and independent part in this debate.
- If media is going to be a serious watchdog in these fields you need a critical mass of journalists who are familiar with science and interested in its implications. They need to meet regularly and exchange views and learn from each other.
- Until now this yearly meeting among journalists has taken place in the USA, at the yearly AAAS meetings. It is now time to put more effort into yearly European Science Journalists meetings and especially in connection to ESOF (European Science Open Forum).
- This is important, not only for the individual journalist, but from the European perspective as a whole.

CHAPTER 29

I HEARD IT ON THE RADIO!

The role of the radio in science communication

MATTEO MERZAGORA[1], ELISABETTA TOLA[2], MARZIA MAZZONETTO[3]

[1]*Science communication consultant and SISSA – Innovations in the Communication of Science, 2/4 rue Mouton Duvernet, 75014 Paris, France, Tel +33-1-45430386, Email matteo@merzagora.net*
[2]*Formicablu and SISSA – Innovations in the Communication of Science, Via sant'anna 16/2, 40128 Bologna, Italy, Tel +39-051-6325626, Email eli@formicablu.it*
[3]*SISSA Medialab and SISSA – Innovations in the Communication of Science, Via Lionello Stock 2/2, 34135 Trieste, Italy, Tel +39-040-3787641, Email marzia@medialab.sissa.it*

Abstract: Radio accompanies our lives. Whether we look for plain information, cultural stimuli, hints for a discussion, campaign arguments, or just fun and entertainment, we will find a radio station which satisfies our needs. New forms of dissemination of audio contents (web radios, streaming audio, podcasting, etc) have further increased the capacity of the radio to fill niches, in a cheap, open, and friendly manner. In many of these niches, we do find science programmes or science topics. The radio session at the CER 2005 conference explored innovative ways of bringing scientific research to our ears, and gave rise to some fundamental questions: Are the sound waves the best place to carve a mature relationship between science and society? Are the innovative audio media well exploited for science communication? How could we promote the presence of science in local, commercial or entertainment radios? What makes a good science story for the radio? Is language a real barrier to communicating European research? What are the "parameters of satisfaction" for a scientist participating in a radio programme?

Keywords: Science in radio broadcasting, Radio, Science communication

1. INTRODUCTION

A session devoted to the radio medium in a conference on the communication of scientific research is not obvious. Until very recently, radio has not been considered a key medium for science communication. At least in theory: because in practice, many radio programmes of extremely high quality are devoted to scientific content in most European countries.

M. Claessens (ed.), Communicating European Research 2005, 169–174.

Literature which specifically discussesthe role of the radio in science communication is also scanty. Most of the existing works consist of proceedings from congresses in which radio practitioners were asked to present their points of view (Dehan 1987, Evans 1993, Williams 1993, Cohen 2001, Mitchell 2003, Jergovic 2004, Coyaud 2004, Redfern 2004, Airhart 2004); radio studies in which science programmes represent good case studies (Fairclough, 1995); historical perspectives (Douglas 1999, Lafollette 2002); and a few analyses of specific topics (Merzagora and Coyaud, 2002; Tola 2004, Napolano 2004, Thakar, 2004, Barone, 2004). A book specifically devoted to the role of the radio in science communication was published only in 2005. (Mazzonetto et al., 2005). Curiously enough, the scientific discourse was a key element in early radio studies: back in 1932 an illuminating text by Walter Benjamin set the stage for most of the discussions currently taking place in science communication studies about its evolution from a diffusive model to a dialogue model. It was the radio – in Benjamin's mind – that was able to open up a new opportunity for science popularisation: 'a form of popularisation that not only mobilizes science towards the public, but also the public towards science [...], and acts on science itself" (Benjamin, 1971).

A session devoted to science in radio broadcasting was thus particularly welcome, and we thank the organisers for inviting us to produce it. We considered this as one key step in a series of activities initiated by the Scirab project – Science in Radio Broadcasting (see below) -, which included the establishment of a network of science radio journalists, a series of workshops and meetings on the role of the radio in science communication organised in the year 2004, a survey of the state of the art and a map of science radio programmes in Europe. The results of the survey and the report of the SCIRAB meetings can be found on the website www.scienceonair.com. A book (*Science in radio broadcasting*) was published as a result of these activities by the authors of this paper (Mazzonctto et al., 2005). It is available for open access in pdf format at www.polimetrica.it, where it may also be purchased in printed format (we invite institutions to buy the printed copy to support the choice of publishing it for open access).

We are confident that this is only a start, and further activities will follow keeping the spotlight well focused on the radio as a key medium for science communication.

2. THE SCIRAB PROJECT

Scirab – Science in Radio Broadcasting – was promoted to share experiences and best practices among science radio programmes in Europe, and to analyse the role of the radio medium from a science communication perspective. The project involved three partners: Sissa in Italy (Innovations in the Communication of Science group), Imperial College, London, UK (Science Communication Group), and Radio Romania, the Romanian State's broadcaster. The project was funded under the Science and Society programme of the EU Sixth Framework Programme for research (2002–2006). One of the main objectives was to conduct a survey of science radio programmes in Europe and organise three international meetings.

The survey was carried out in 2004. A map of 75 science radio programmes being broadcasted in Europe was drawn, and 40 programmes from 16 countries provided answers to a detailed questionnaire.

This survey uncovered a much richer landscape than previously expected: by travelling through Europe, it is indeed possible to listen to a science radio programme at any time of the day, any day of the week. Each of the programmes shows a different approach to science and science communication, with new ideas sparking everywhere. We strongly suggest everyone interested in science radio broadcasting to take a trip along the map we have tried to draw (www.scienceonair.org: the web offers a great opportunity for this): creativity is a quality that is not lacking among science radio journalists!

About 50 science radio journalists and scientists from 16 countries actively participated in the Scirab meetings.

A first workshop was held in Bucharest on the premises of Radio Romania. It was specifically devoted to "Challenges and opportunities in Radio Science Journalism in Eastern Europe". The aim was to explore similarities and differences in the ways of conceiving science communication in the new Member States with respect to the old European countries.

A second meeting was held in Trieste. More than 30 science radio journalists from 16 different countries met, together with scientists who participate in radio programmes and media researchers specialised in radio studies. The meeting was devoted to "The role of the radio in science communication". A panorama session allowed all of us to present our programmes, to discuss choices and to exchange good ideas, while a series of lectures highlighted some key points, from the relationship of the science journalists with the editors/broadcasters to the opportunities of international co-productions, from language issues to the role of the Internet.

The third meeting was held in London. It was entitled "Are scientists heard on the radio?", as it focused on the role of the expert in science radio programmes. An enlightening exchange of points of view regarding challenges of science communication among scientists and journalists took place on the premises of the historical BBC Bush house.Results of the survey and the proceedings of the Scirab meetings are available on-line at www.scienceonair.org, and in Mazzonetto et al, 2005.

3. THE ROLE OF THE RADIO IN SCIENCE COMMUNICATION

The debates on most current controversial issues, from climate change to biotechnology to nanotechnology, "cannot really avoid developing in conformity to the principles of the social environment, rather than those of the scientific world where it was conceived" (Ziman, 2000). The dialogue between science and society, which is advocated in a variety of institutional documents, needs opportunities to be experimented in practice, in forms that can provide beneficial outcome for science, for society and for the media, taken alone or in their mutual interactions. Indeed, radio appears as the media space where science, in its current evolution toward a post-academic era, can best find an opportunity to meet the public.

The capability of upstreaming controversies (that is, to encourage a dialogue "at an early stage in the process of new developments, when possibilities, problems, ethical issues and risks are being projected rather than at a point when public anxieties have become polarised" (Whitmarsh et al., 2005) is presently a major challenge of science communication. We wish, and we believe, that radio will be able to meet this challenge, but we will need to clarify what its limitations and opportunities are, starting from the role that the expert can assume in a science radio programme. By being present with their own voice, and at the same time not being obliged by technical constraints to follow a strict script, scientists on the radio can be both accountable and express themselves clearly: in other words, they can find on air the path through which to renovate their pact with society at large.

If, as several studies have recently pointed out, today's challenge for science communication is promoting public participation and developing an informed, constructive and democratic criticism of scientific research, then radio can be of great utility.

Compared to other media, radio can easily establish contact between the listener's personal experience (both practical and cultural) and the world of scientific knowledge. Unlike written media, radio allows the audience to hear the voice of the protagonists, thus creating a more intimate connection with the world of science and it is able to show the motivations behind the scientists' work. Compared to television, radio is a lighter medium, with a more relaxed and reflective tempo, and the barrier separating the listener from journalists and scientists is less impenetrable. The stereotyped and standardized roles so frequent on television could not work on radio: we create our own opinion of the scientist we are listening to, basing it on his/her style and on what he/she is saying. This is why people trust radio so much: scientists who are being interviewed are personally responsible for their own image and can be judged by the public. They cannot hide behind their expertise and, at the same time, they can use their knowledge to convince the public of the interest, relevance and usefulness of their work.

In brief, radio is so effective in communicating science because it relies on a very familiar mode of communication: conversation.

As both a prerequisite and a consequence of this, it is important to critically address the issue of what role the journalist has in a science radio programme. As stated by Mitchell (2003): "The public increasingly takes science for the indefinite, political and often messy business that it actually is. Accordingly, we in the media can no longer get away with reporting science in the naïve celebratory way we did 30 years ago". One of the main issues is where a science journalist should place him/herself along the channel that connects the scientist and the public: the choice whether to be "close to the scientist" or "close to the listener" is determinant in the type of communication that is established, and while it can often be hidden or implicit in written journalism, it needs to be directly addressed when talking on the radio.

4. RADIO AT CER 2005

The radio session at the Communicating European Research conference was structured as a series of short and lively presentations, spanning many different approaches to the topic. Elisabetta Tola (Formicablu srl and RAI – Radio 3, Italy) set the scene and introduced the Scirab network. Jan-Olov Johansson (Swedish Radio) presented the successful experience of science programming in Sweden, testified by listeners' surveys and by the fact that "Science News" is currently the most downloaded podcasting of the radio station. Chris Smith (The Naked Scientists, UK) showed how combining radio, web, podcasting and a fresh and lively approach can conquer new audiences. Pedro Lima (Association Marseille Science and Radio Grenouille, France) concentrated on the experience of bringing science in a local radio, and from there to a network of local radios, rarely featuring science programmes; from local to global, Laura Durnford (Radio Netherlands) presented the challenge of bringing a programme like "The research file" to a global audience, while Joanne Carpenter (Panos London) underlined the potential of the radio for presenting and discussing research issues in developing countries.

It is obviously the combination of many different styles and approaches, and the opportunity of experimenting at relatively low cost, that will set the future of science on air. We hope that many more ideas will emerge, on air and in future conferences and meetings.

REFERENCES

Airhart M (2004) Earth and sky: Some challenges in communicating scientific research on the radio. In: Chittenden D, Farmelo G, Lewenstein BV (eds) Creating connections, Altamira press, Walnut Creek, pp:47–60

Barone LT (2004) I cantastorie della scienza. Elogio della variet‡ nel panorama radiofonico Europeo. In: Pitrelli N, e Sturloni G (eds) La comunicazione della scienza, Zadigroma, pp:289–298

Benjamin W (1971) Drei Hormodelle. Suhrkamp Verlag, Frankfurt am Main

Cohen D (2001) Presenting a radio science program: Engaging the public interest. In: Stocklmayer S, Michael M Gore, Chris B (eds) Science communication in theory and practice. Kluwer Academic Publishers

Coyaud S (2004) Retorica, impatto cognitivo e impatto emotivo: Il respiro della scienza e quello della radio. In: Pitrelli N, e Sturloni G (eds) La comunicazione della scienza, Zadigroma, pp299–302

Deehan G (1987) Science broadcasting – its role and impact. In: Wiley (ed) Communicating science to the public. Ciba Foundation Conference, p:88–99

Douglas S (1999) Listening in: Radio and the American imagination, from Amos 'n' Andy and Edward R Murrow to Wolfman Jack and Howard Stern. Crown Publishing Group

Evans P (1993) Radio science: Form and function. In: The role of the media in science communication, Ciba Foundation discussion meeting, Stockholm, pp:123–137

Fairclough N (1995) Media discourse. Arnold Eds., London.

Ferrante A (2004) Utilizzi degli archivi per nuove forme di comunicazione radiofonica. In: Pitrelli N, e Sturloni G (eds) La comunicazione della scienza, Zadigroma, pp303–306

Jergovic B (2004) Choices that make radio science stories, Jcom 3 (4), http://jcom.sissa.it/comment/com030403.pdf.

LaFollette MC (2002) A survey of science content in US radio broadcasting, 1920s through 1940s, Science Communication 24, p4

Mazzonetto M, Merzagora M, Tola E (2005) Science in radio broadcasting: The role of the radio in science communication, Polimetrica

Merzagora M (2004) Science on Air: The role of the radio in science communication, Jcom 3 (4), http://jcom.sissa.it/comment/com030401.pdf.

Merzagora M and Coyaud S (2002) Conversations between scientists and the public in radio PHONE-INS: An experimental approach to analyse public perception of science. Jcom, 1 (1). http://jcom.sissa.it/article/art010102.pdf

Mitchell G (2003) Science for a sceptical audience, Institution of Electrical Engineers, 2003. http://www.iee.org/OnComms/Circuit/benefits/science_scepticalaudience.cfm.

Napolano V (2004) La comunicazione della salute mentale: Un corpo a corpo radiofonico. In: Pitrelli N, e Sturloni G (eds) La comunicazione della scienza, Zadigroma, pp307–314

Redfern M (2004) BBC radio science: Challenges and opportunities, Jcom 3 (4), http://jcom.sissa.it/comment/com030404.pdf

Thakar B and Kothari A (2004) Innovations to enrich science communication through radio, Jcom 3 (4), http://jcom.sissa.it/comment/com030405.pdf

Tola E (2004) Verba manent e remediation: Esperienze di convergenza fra radio e web. In: Pitrelli N, e Sturloni G (eds) La comunicazione della scienza, Zadigroma, pp315–324

Williams R (1993) Old blokes in white coats. In: The role of the media in science communication, Ciba Foundation discussion meeting, Stockholm, pp189–200

Wilsdon J, Willis R (2004) See-through science: Why public engagement needs to move upstream, DEMOS, London

Ziman J (2000) Real science. What it is and what it means, Cambridge University press

CHAPTER 30

COMMUNICATING RESEARCH IN DEVELOPING COUNTRIES

JOANNE CARPENTER

Programmes Officer, Panos, 9 White Lion Street, London N1 9PD, United Kingdom, Tel +44 207-2397635, Email joanne carpenter@panos.org.uk.

Abstract: With regards to science communication, it is useful to ask 'Why use radio, and what for?' We have to look at communicating through radio in terms of how it can enable ordinary people to make sense of their lives, to engage as citizens in their society, to get information, and above all, to express a voice. It is important for researchers to be clear about who they want to talk to through radio. It might be the people who will use your research, or constituencies that are the subject of your research, or the decision makers who can take up your recommendations. The RELAY pilot project has found that radio, alongside other media, can act as a platform for debate around academic research reaching all these groups

Keywords: Science in radio broadcasting, Radio, Science communication, Communicating research in developing countries

1. PANOS AND COMMUNICATION

Panos is a network of institutes working in Europe, Africa, South Asia and the Caribbean on the role of communication in achieving development. We believe that communication must reach everyone in society . . . and everyone must be able to communicate.

As well as working with radio, we work with TV, print and online media, civil society and governments to promote inclusive and informed debate around important development issues: globalisation, environment, HIV/AIDS, reproductive health, information society and conflict. We do this through different means: promoting discussion in the media, building the capacity of journalists, lobbying editors to cover these issues and advocating for appropriate government legislation to be passed. Communication among individuals, businesses and communities, between civil society organisations and their constituencies, and from government to people and vice versa, all contribute to the

M. Claessens (ed.), Communicating European Research 2005, 175–179.

growth of equitable societies and can address poverty. Communication plays a key role in the following areas:

- Effective provision and take-up of services such as health and education
- Economic activity, trade and economic growth
- Empowerment of individuals and communities
- Dialogue between people and decision-makers, including consultation and participation in development strategies and democratic political processes
- Transparency and accountability of governments, businesses and other influential groups
- A "public space" for discussion of social, cultural, economic and political issues.

2. RELAY 2004–5 – WHY ARE WE COMMUNICATING RESEARCH, AND WHO ARE WE COMMUNICATING IT TO?

For the past year, Panos has been running RELAY as a pilot project, working on the premise that academic research on critical global development issues needs to go beyond research communities and reach ordinary people in developing countries – the very people who will be the most affected by policies based on research findings.

RELAY aims to strengthen the role of media in effective communication of academic development research:

- To disseminate findings and recommendations more broadly
- To stimulate civil society debate and action around them
- To be accountable to those affected by research
- To generate debate which can impact on policy.

To date, RELAY has supported the production of radio features by journalists in East Africa, Southern Africa and South Asia, produced radio discussion programmes and provided forums for journalists and researchers to discuss opportunities and obstacles in working together.

3. WHY RADIO, WHAT FOR?

It is useful to ask 'Why use radio, and what for?' We have to look at communicating research (CR) through radio in terms of how it can enable ordinary people to make sense of their lives, to engage as citizens in their society, to get information, and above all, to express a voice. It is important for researchers to be clear about who they want to talk to through radio. It might be the people who will use your research, or constituencies that are the subject of your research, or the decision makers who can take up your recommendations. RELAY has found that radio, alongside other media, can act as a platform for debate around academic research reaching all these groups.

Radio is a critical medium for two-way communication. A revolution has taken place in terms of the amazing de-concentration of ownership, the extraordinary liberalisation of broadcasting in particular, and of media in general that has taken place worldwide, and particularly in the developing world. There are almost three billion people on this planet who live on less than 2 dollars a day. Ten years ago, most of the information they had access to came from governments. We have seen this extraordinary blossoming, and mushrooming of radio in particular in developing countries. 15 years ago, there were 10 independent radio stations in all of Sub-Saharan Africa and today there are thousands. In Uganda there are just short of 80 radio stations, most of them independent and commercial – ten years ago, there were only two. There is a huge wealth of chat shows, phone-ins and discussion programmes being broadcast that academic research can engage with. *Radio enables academic researchers to be accountable to people living in developing countries that are normally excluded from debates – those people whose needs and rights are often ignored by society and government policy.*

4. WAYS RADIO CAN BE USED

Here are some questions researchers should ask themselves about CR. There are choices to be made about how to engage with radio – are you seeking to communicate a particular message or particular information to which you want listeners to respond in a certain way? Or are you simply putting your research out in the public domain to be debated?

Radio enables researchers to be accountable to people living in developing countries, those that are normally excluded from debates, those who have no voice and those whose needs and rights are ignored by society and government policy.

By engaging with the feedback that radio programmes can elicit, research agendas can be reinvigorated with issues that are really relevant to the majority of ordinary people living in developing countries; researchers can get out from behind the desk and become part of inclusive policy influencing processes where voices of groups in society that are normally powerless and cut off from public debate are listened to.

5. RELAY EXPERIENCE IN USING RADIO TO PROMOTE DEBATE

In the past year, RELAY has promoted debate around specific pieces of academic research through one-hour radio discussion programmes produced by commercial FM stations in rural Zambia (on genetically modified – Bt – cotton) and in Accra, Ghana (on educational reform). In both cases, people who would not usually have access to these debates phoned in to voice their concerns, and in the case of Ghana, a question was asked in Parliament as a result of the programme. RELAY has also commissioned 24 journalists from developing countries to produce radio features which scrutinise a particular piece of research, asking questions about the research findings on the ground and getting ordinary people's responses to the issues raised by the research. All RELAY features put a local issue in a global context – ensuring that the features are relevant to radio stations all over the world. The five-minute features are used in a variety of ways, unedited in their entirety, as part of a broader programme, translated, or used for training purposes by editors and producers. They can be accessed from our website, but problems with internet access in the developing world mean that we do not rely on this method. We also send out CDs of a series of features to our network of 600 radio stations who have signed up to be members.

RELAY has also identified a huge role to play in increasing the capacity of media and researchers in developing countries to engage with and understand how each other work. RELAY forums in Zambia, Uganda and Sri Lanka brought together journalists and researchers. They demonstrated that there is real potential for collaboration between researchers and the media in developing countries, but also highlighted a mutual suspicion. In many ways, the two groups want to achieve the same thing, not least in communicating important new academic research evidence and recommendations to the public. But obstacles do exist . . .

The situation for radio journalists RELAY worked with is very different to the experience of the journalists working in Europe. Most journalists here and on the panel specialise in science, but in the African and Asian countries that RELAY works in, young journalists are often starting from a lower skills base. The profession is low status, and journalism training schools do not tend to encourage specialisation. Presenters working in provincial radio stations are often high school graduates with no journalistic training at all. By working on RELAY features, journalists in developing countries have found that covering academic research added credibility to their work, brings them to the attention of editors and increases their confidence in putting questions to 'experts'. Radio journalists that RELAY has worked with often think they are not qualified to scrutinise research, they feel they do not understand research ethics and are put off by unfamiliar terminology and statistics. In many cases, local researchers are unwilling to be interviewed by the journalists and to explain their research. The RELAY experience is that researchers need to go more than half way to meet journalists, making themselves available for interview and answering telephone queries.

A number of recurring themes emerged during the forums:

- Most of the research done is donor-driven, meaning that issues that affect ordinary Ugandans, Zambians and Sri Lankans are not researched – researchers can benefit from feedback into their research agendas in mitigating against this.
- The commercial radio and print media prefer 'sensational' stories which sell.
- Research institutes are also subject to commercial pressures, making them unduly concerned with controlling their image and coverage of their research output.
- In Uganda, the prevailing culture undermines the spirit of research, investigation and scrutiny. This may also be a factor in other developing countries.
- Not all research findings are 'news' – this makes it difficult to cover them.
- Journalism training institutions do not promote specialisation in a particular topic such as health, education or trade.

In all three countries where RELAY hosted a forum, both parties acknowledged that it was important to actively seek each other out, and not to sit and wait to be contacted. They also identified a need to build mutual respect and to cultivate strong working relationships.

In the next phase of the programme, RELAY will be working to create pools of young journalists who are familiar with the use of academic research in their journalism from an early stage in their careers. RELAY capacity building activities will encourage specialisation and work with radio editors to promote coverage of important development themes. In working with researchers in developing countries, RELAY will seek to change attitudes about engaging with the media, and where possible, work together with research institutes in developing research communication strategies which include the media. The potential for two-way communication between ordinary people in developing countries and researchers will continue through radio discussion programmes but new activities will include live debates bringing Northern and Southern researchers together to discuss nationally relevant issues, and making programmes where researchers take their research findings back to the communities that they researched.

CHAPTER 31

GETTING R&D RESULTS INTO THE PRESS

Information society technology results

TARA MORRIS[1], GARETH HARDING[2], LAURA MILES[3], ERIC CHREIKI[4]

[1]*GOPA-Cartermill, 45, Rue de Trèves, 1040 Brussels, Belgium, Tel +32 2 286.19.85, Email tmorris@gopa-cartermill.com*
[2]*United Press International, Brussels International Press Centre, Residence Palace, Rue de la Loi 155, 1040 Brussels, Belgium, Tel +32-2-2302174, Email:garethharding@skynet.be*
[3]*AlphaGalileo Foundation, 175-185 Grays Inn Road, London WC1X 8UP, United Kingdom, Tel +44-20-78120670, Email laura.miles@alphagalileo.org*
[4]*Innov-Europe, Espace Lumière 8, 57 boulevard de la République, 78400 Chatou France, Tel +33-1-30719728, Email ech@innov-europe.com*

Abstract: The breadth of media channels covering research grows daily. As an important channel for research projects to promote their results and work in progress, an attempt was made to demystify the workings of the media and provide practical guidance on how to get research results into the press

Keywords: Results, Wires, Syndication, Press Release, Media Strategy

1. INTRODUCTION

The growth of press publications interested in covering research is an important channel for projects to promote their results and work in progress. For many scientists this can be very challenging and often the benefits of their research remain hidden due to lack of communication expertise. Journalists too find it difficult to trace projects' results in the maze of information.

But where should projects start? Should the focus be on major European titles or smaller, specialist publications? What makes a press release that will appeal to journalists and be noticed in the crowd? And what is the best way to approach media professionals?

M. Claessens (ed.), Communicating European Research 2005, 181–186.

The session aimed to answer these questions, providing researchers with advice from experienced journalists and European press services on professional approaches to ensure news is taken up by the media.

Around 75 people attended the workshop representing research projects and their communications officers, European Commission staff and a range of support organisations interested in research communication.

2. PROMOTING EU-FUNDED ICT PROJECTS, T. MORRIS, EDITOR-IN-CHIEF, IST RESULTS

IST Results[1] is an online news service launched in March 2003. It raises the visibility of Information Society Technologies (IST)-funded research results, supports projects' market access, and raises the awareness of the IST programme and its activities.

The service offers feature articles on specific IST projects complemented by regular news in brief on IST-related topics, an IST projects events calendar and comprehensive links.

A dedicated press desk and investors room offer filtered entry points while the weekly e-bulletin, customisable email alerts, and syndicated content via press wires and RSS (Really Simple Syndication) news feeds push the content out to users.

The editorial process follows a monthly thematic calendar focusing on specific topics as well as reporting on current developments in all IST areas.

IST Results has found that a particularly successful approach in raising the visibility of IST project results has been the syndication of IST Results feature articles to various press and wire services such as PRWeb, CORDIS Wire, AlphaGalileo and EurekAlert! On average IST Results syndicates eight articles per month to wires.

The use of wires has led to articles being taken up by various mainstream media publications including the Financial Times, Wired, VNUnet, New Scientist and The Engineer. Also, many other organisations have republished IST Results in their own online newsletters and web pages, providing the service with many opportunities to reach audiences with a particular technology or market interest or those based in a specific country or region.

IST Results content is available to the media and other users via a range of RSS feeds. IST Results also actively promotes project results through a dedicated GoogleAds campaign that generates over 9 000 clicks per month.

All told, this has led to over 1300 projects being covered in some 800+ published features and the 100 weekly news articles. Overall, the approach has generated very positive feedback and appreciation from featured projects and users alike.

---

[1] http://istresults.cordis.europa.eu.int/index.cfm

3. GETTING RESEARCH NEWS INTO THE PRESS, G. HARDING, CHIEF EUROPEAN CORRESPONDENT, UPI BRUSSELS

A series of tips and observations on dealing with the press was presented.

Too many organisations send press releases to say that something has been launched or something has happened. This is not news. They need to pick out what is really newsworthy. Ask what difference does this make? Why should anyone be interested? Journalists are taught not just to report the what, when, why, where and who but also the 'what difference will it make?'. Interpret the facts and focus on the why and what difference it will make in press releases.

Journalists receive a lot of the aforesaid press releases. Aim for one page (maximum two) and send them by email not by fax. However, email is easier to delete so make sure the title and first sentences invite them to read on. Make quotes quotable and ensure the press contact details are up-to-date and the person is available. Use personal mobile numbers not office numbers.

Typically journalists get around 200 emails daily so they have to process them quickly. A large percentage are not news so are deleted straightaway.

Do not keep sending press releases on a regular basis with updates. This can result in journalists deleting them as soon as they see the company name.

Do not hassle journalists. If they like the story, they will get back to you very quickly. They are not looking to build a relationship. They need to stay independent.

For press conferences, rehearse properly, and anticipate questions and answers. Know the lines. Have quotes ready. Do not bother with fancy lunches or free wine – journalists are not so interested in this. Keep presentations short, say around 5–10 minutes.

Respect deadlines. Take time to understand the media timetables for the different publications as these differ for the daily, weekly or monthly press.

At interviews, set the rules in advance. Journalists will assume everything is on record unless told otherwise. Specify when it is off the record, where the quote may be used but indirectly, not naming the person or the organisationspecifically or when the information is to be used as background (no quotes used).

If the journalist does not use the material this time, there may be another opportunity but not if he/she is hassled about it not appearing. And never lie – it always comes out! Journalists tell of examples of reporting on so-called unique breakthroughs where it turned out there were other similar developments already underway.

Do not underestimate the journalist's knowledge. Many lack scientific knowledge but this can be useful as it forces scientists to explain simply what their development is all about. Assume ignorance in the target audience and aim at the lowest common denominator as the general press in particular is aimed at the public, not experts.

Journalists prefer to publish without getting your final approval, as they often have to wait too long for scientists. Researchers should not be afraid or paranoid and should learn to trust journalists.

4. ALPHAGALILEO, THE RESEARCH PRESS SERVICE, L.MILES, OPERATIONS DIRECTOR

AlphaGalileo[2] was set up in 1998 to highlight the significance of European research via the press and media. It takes press releases about research developments from over 2 000 contributors and distributes them to more than 5 300 journalists in some 85 countries. Contributors pay an annual subscription to use the service.

AlphaGalileo keeps regular contact with existing users to encourage most effective service. In each country it targets research organisations to register and regularly add content to AlphaGalileo and media organisations to encourage journalists to register and actively use it. It also offers expertise to both users and non-users on best practice in communicating research.

The service is continually updated, and the content and users are checked by experts. It is quick and easy to post content such as releases, events, books, broadcast items and background material in any European language.

For contributors there is statistical feedback and they can embargo content. The media benefits from an experts' database, customised email alerts and news feeds, language settings and an address book of contributors.

AlphaGalileo is also one of 24 partners and supporters of the Communiqué initiative, a vehicle to help progress the evaluation and subsequent creation of a European research media service. Communiqué seeks to work with all interested bodies and individuals to achieve the establishment of a programme to deliver the functions recommended for the service. Communiqué is not the media service, but an impartial initiative to ensure that any media service that is created is relevant and practical.

5. BUILDING AN EFFECTIVE MEDIA STRATEGY, E. CHREIKI, EDITOR INNOV-EUROPE, MEDIA STRATEGY EXPERT

Focusing on the importance of planning media relationships and using the right communication tools at each stage of the strategy, a comparison was made between the 'Go to Market' approach to R&D results' communication and exploitation, and the traditional 'So What' approach.

Typically, projects only communicate at particular stages of their work, publishing their results in a journal and, if they develop a product, publishing a press release to announce its launch. If there is no relationship with the target audience, they may not be receptive to a press release and projects may have lost an opportunity to make a market impact.

A more effective approach is to adopt a market-focused strategy that incorporates communication in all phases with relevant stakeholders.

---

[2] http://www.communique-initiative.org/

In raising funds, projects should aim to alert investors and institutions and stress the market needs and technical relevance. This can be achieved by means of conferences, lectures, interviews and scientific publications.

By preparing industry and partners the needs and potential market should be emphasized. This calls for attracting potential partners and the establishment of new links.

As a project, protect the IP rights and concepts, evaluate the market, and get feedback. This involves market studies, reports, white papers branding, partnerships, awards and regular 1-to-1 contact with selected journalists.

The next phase involves alerting and preparing the channel. This serves to raise awareness of the first buyers and get the support of influencers. For this projects need to consider press briefings, regular releases, market study results, fairs, customer cases, user communities and websites.

The final milestone consists of preparing customers to create or enforce customer needs and attract potential partners. The strategy here could consist of using focus groups, customer studies, design and market tests and press events with partners. Using the right tools at the right phase is also essential.

Although both have similar overall costs, the 'Go-To-Market' model spreads the cost over time, resulting in on-time delivery of the media message. Compare this to the 'So What' large lump sum model that may result in the late delivery of the media message. As a result, the 'So What' approach often has a poor impact and low benefits with a better impact and higher return achieved by the 'Go-To-Market' approach.

Innov-Europe[3] provides guidelines on costs and return on investment of tools for research, and examples of innovation marketing best practice.

6. QUESTIONS AND ANSWERS SESSION

6.1. What Sources do Journalists Use to Find Leads?

Often journalists do not use scientific news feeds as it is not their core area of interest. Instead, they use political news wires.

Try to get a mention in some of the magazines that journalists read to have indirect contact. Key publications really count – journalists read Reuters whose news is picked up by Associated Press who then feed the Financial Times which is also read. It can be pretty incestuous.

Remember there are many types of publication as well as the main business dailies – monthlies, features, supplements, men and women's magazines.

Features rather than news may be better for research communications as there is less need to reduce the text to a minimum or to dumb it down.

[3] http://www.innov-europe.com/

Look at press releases as a sales pitch or advert and put the rest of the information elsewhere, such as in notes for editors or links.

6.2. Is There a Trend Towards Reporting What Sells and Not Science?

Publishers are private sector organisations. It would be good to be high minded but publishing equals money.

Today there are fewer special correspondents in the press. As a result there are fewer opportunities for coverage and so hot topics win out.

In general there is a tendency to dumb down. Take the over-simplified example of a choice between the latest chemicals directive and a star's new boyfriend. There is a clash between the important and the interesting. We know the directive is more important but it is more likely that the boyfriend article will be more prominent in all but the most serious publications.

Maybe one solution could be to include market or society data with press releases to give the reader/editor more populist content.

Part of the problem is often the low level of education of readers. Only 5% of people fully understand the content of the daily news. Science is often the second topic of interest in newspapers after politics. However, science is complicated and so readers need more time to digest it.

6.3. Other Tips

Journalists need a daily news feed – it is quite formulaic even for scientific news. So get in touch with journalists early to share the vision and prepare them for results later.

There is often only one person to cover many areas. So coverage of research is only a small part of these correspondents' work.

Assume all journalists are generalists and start with a clean slate when describing what has been achieved. 'New' means it is new for journalists too.

Journalists need good sources – if they think they will get a good story, they will be interested. They often need quick quotes regarding European issues and will re-use groups or individuals that are trusted and who have been useful in the past. If they do not get the right experts or quotes, they will not use that source again.

Expert contacts mentioned in press releases need to be reliable and available when the press release/news story goes out.

It may be possible to get research news in on the back of other breaking stories. If this is the case, get your people and sources ready for quotes.

Has there been any market research or data? This is very interesting for journalists. This sort of data can be added to the Notes for Editors along with background data on the organisation.

CHAPTER 32

TOWARDS MORE RESPONSIBILITY IN COMMUNICATING SCIENCE

BLANKA JERGOVIĆ

Croatian Radio, Prisavlje 3, 10 000 Zagreb, Croatia, Tel +385-1-2395937, Email b.jergovic@hrt.hr

Abstract: Public communication of science today is more and more characterized by the combination of scientific and social priorities. Complex scientific information, application of scientific knowledge and its implication on society need to be communicated to the public within the new paradigm of media communication of science. This is impossible without science journalists but requires a new prototype of science journalist. At the same time media performance of science is a journalist-scientist product. The scientific community is not only the main and often the only source of information. It sets the agenda for the media communication of science. Therefore, the responsibility for the quality of that communication is shared between journalists and scientists

Keywords: Science journalism, Media, Communication, Scientists, Journalists

1. INTRODUCTION

Is science journalism necessary? Unlike previously, today's science makes the headlines in the radio or reaches the front pages of the newspaper mostly in times of uncertainty and crisis. At this point, the scientific topic is often taken over by journalists from the news desk who are responsible for an effective broadcast. Does this mean that science journalism is a luxury? Science journalists not only have the function of news reporting but also act as knowledge interpreter between scientists and the public. Hence, science journalists generally need more resources to process scientific news in order to reach a broader audience than their colleagues from other news departments.

At times of high scientific media coverage, currently exemplified with the bird flu, the quality of science news reporting often declines with an increased demand

M. Claessens (ed.), Communicating European Research 2005, 187–190.

for clarification. As scientific topics are increasingly on the political and social agenda, they are more and more subjected to controversies and debates. Moreover, they can serve as paroles for political parties and can be used to shape a politician's profile and might be exploited by various interest groups. In times that call for the need for "conversation and participation" rather than "sensation and polemic", science journalists should demonstrate their need in such a manner that society will not even question whether science journalism is a luxury or not.

2. MEDIA AND ITS COMMUNICATIONAL IMPACT

Media is the most common source of information about science and technology for the public. According to latest Eurobarometer data and, for example, the results from NSF Science and Engineering Indicators 2004, the majority of the public obtains the information about science from television. Media interest in science is reflected in the presence of science in the non-scientific newspaper and magazines. Out of all articles published, 2% in the USA and approximately 5% in the UK is about science. In Greece it is 1.5 to 2.5%, in two most popular dailies in Croatia 3.5% and in South Africa 1.8%. Studies have shown that the most frequent readers of media science are scientists. Scientists represent a special readership; their interaction with science journalists and their genuine expectations strongly affect media coverage of science. Both scientists and science journalists historically cultivate special relations that also influence the way in which science news is reported. On the contrary, in a world of huge scientific output, the media can also be a guideline for scientists and even set the agenda for science. For example, a paper published in the highly quoted *New England Journal of Medicine* is three times more likely to be cited in the scientific literature if it has first been mentioned in *The New York Times* (Phillips, 1991). We can therefore conclude that science journalism is crucial in the public communication of science.

The perception of its communicational impact, however, differs according to if we were scientists or journalists. Scientists often complain that journalists do not describe "real" scientific work, but rather present an exaggerated or even alarmist, negative image of science. But the contrary is also true: there is a tendency in the press to report about science in a very positive, even ecstatic way (Schnabel, 2003). Scientists emphasize the need for more accuracy in science journalism as a way to diminish misinterpretation of scientific information (Bucchi, 2004). Their main criticism is that details are omitted (Dunwoody, 1986), and some even would like to see scientific methods reported. Another comment often heard by scientists is the oversimplification of scientific issues and the tendency to sensationalize pure facts pulled out of context.

From the journalist's point of view, science journalism requires more effort than any other news reporting and is often combined with the feeling that their work is less appreciated. Science journalism, however, requires a certain scientific background in order to transform scientific complexity into a story, or at least into comprehensible information. Writing about science is time consuming and often science journalists stay with features, investigative and explanatory articles with unpredictable outcome.

Science journalism is an everyday fight with the editors to convince them to publish news of science and technology (see e.g. Gregory and Miller, 2000). In broadcast media, science stories are mainly placed at the end of the news and are more likely to be shifted or not published at all. Even worse, the science section often succumbs to budget cuts. As we can see, scientists and journalists (and even more media editors) follow quite different routes of interest. Nevertheless, media performance of science is a journalist-scientist joint product, and scientists are more than just the source of information.

3. TOWARDS A NEW PARADIGM OF MEDIA SCIENCE

Public communication of science today is characterized by the combination of scientific and social priorities that requires a new prototype of journalist. This could mean that we need to overcome the difference between professionally skilled news journalists with their ability to arouse interest in the reader by the rules of the media market, and science journalists with their profound insight into science. This calls for new models of education for science journalists. Preliminary data show that currently there are 22 mainly academic training opportunities for science journalists in Europe with an average of 2 per country. The countries with most training opportunities are Germany (with more than 30 in science journalism and communication – 4 are academic), France (5) and Sweden (3)[1]. The duration ranges from one to three days up to one semester or more.

No specific training in science journalism exists in Russia, Slovenia and Denmark. In some countries, such as Hungary, the national association of science journalists developed their own training programme. In Russia, and more specifically at the University of Moscow, some of the journalists of the Russian Science Journalist's Association give schoolmasters classes in science journalism. These lessons are very short, i.e. maybe one or two hours. In Croatia, there is a one-week science communication summer school aimed at postgraduate students and journalists. There is only one private training course in science journalism in France, one in Switzerland and two in Hungary (the data is derived from the French Association of Science Journalists which conducted a survey among national science journalists' associations belonging to the EUSJA in May 2005 including answers from twelve countries: France, Sweden, Ireland, Denmark, Germany, Switzerland, Russia, Slovenia, Hungary, Belgium, Croatia, Austria). The new models of education for science journalists have to face two main demands set by the complex science-society relations: more sources of information and more criticism.

4. SHARED CULTURE VERSUS CRITICISM?

We all know that the media need reliable sources of information, reachable experts who have good communication skills and topics which would attract the highest number of readers. Now, the problem is that scientists and journalists belong to the

[1] This does not include data from United Kingdom, which would drastically increase the number of courses.

«shared culture» that creates a sort of «outsiders and insiders» in the media, and which results in a limited and smaller number of scientists quoted (which can also be the consequence of limited time and resources to make more extended research, and the «we also have a scientist working on a similar topic, let's ask him» approach). On the other side there are omni-present experts who circulate in the media. The situation can be reversed when the impulse comes from the scientific community. Almost 25% of the articles published in the Italian quality daily newspaper *Il Corriere della Sera* in the last 50 years were written by researchers and doctors (*Il Corriere della Sera* has an extremely high number of articles about science: 28% of all articles). Almost 25% of the articles on science in the UK press starts from initiatives by the research community (Bucchi, 2004). If we take into consideration that science journalism tends to be one-sourced and therefore largely depends on the scientific community, especially on scientific magazines like *Nature* and *Science*, (which are the most common sources of media communication of science), the media is to a certain extent controlled by the scientific community who sets the important part of the agenda (de Semir, 1996).

Apart from the existing difficulties to communicate science, there is this huge task to keep pace with the ever-growing scientific knowledge. It is thus not surprising that science journalists rather tend to feature researchers in a story or an interview, or follow the agenda setting of scientific journals. As in arts or literature, criticisms belong to the most important and appreciated part in the media culture. Science as part of our culture has still not reached this level. The daily flood of scientific data and the widening gap between science and the public call for a new generation of science journalists – or science critics? – who are specifically trained to evaluate in an unbiased manner the validity of scientific data. Never before has there been such a need for science journalism unless we deem cultural criticism to be pure luxury.

My thanks go to Thomas Schnyder, science writer, for fruitful discussions and critically reading the manuscript.

REFERENCES

Bucchi M (2004) Science in society, Routledge, London, U.K. and New York, U.S.A

Friedman SM, Dunwoody S, Rogers CL (eds) (1986) Scientists and Journalists, AAAS, USA

Gregory J, Miller S (2000) Science in public, Perseus Publishing, Cambridge, Massachusetts, USA

Phillips DM (1991) Importance of the lay press in the transmission of medical knowledge to the scientific community, New England Journal of Medicine, 11 October:1180–1183

Schnabel U (2003) God's Formula and Devil's Contribution: Science in the Press, Public Understanding of Science, 12(3):255–259

de Semir V (1996) Medicine and the media: What is newsworthy?, The Lancet, 347:1163–66

European Commission, Eurobarometer 55.2 survey and standard report, Europeans, Science and Technology, December, 2001

European Commission, Special Eurobarometer, Europeans, Science and Technology, June, 2005

National Science Foundation, Science and technology: Public attitudes and understanding, (http://www.nsf.gov/statistics/seind04/c7/c7s1.htm)

CHAPTER 33

EUROPEAN MEDIA: TWO CULTURES OF SCIENCE COMMUNICATION

A metropolitan strategy to promote scientific culture at local level

VIOLA EGIKOVA

Science journalist, President of Russian Association for Science Writers and Journalists Intellect, Horoshovskoe shosse 50-98, 123007 Moscow, Russia, Tel +7-495-256 5122, Email egikova@mospravda.ru

Abstract: The situation of Russian science journalism is very special. On the one hand, positive steps have been taken which show that an attempt has been made to create new standards of journalism as it plays a major role in the European community. On the other hand, the reality is rather incompatible with the country which used to be known for many years as the home of intelligent and wel educated persons, with fine examples of science. However this chapter provides several examples which show that Russian science journalists try to change the situation

Keywords: Science journalism and communication, Media

If you ask about science journalism in Russia, you will get very different points of view. One person would say: nowadays media provides the public with plenty of useful and very interesting information. But someone else would say, like Professor Sergey Kapitza, well known in the former Soviet Union as the author of the science popular TV programme "Clear – Unbelievable" (according to the article of Andrey Vaganov in magazine Journalist): "If our times will be ever judged, media should be classified as a criminal institution because journalists manipulate the public's mind. That is a crime and no less!"

What is the truth? Which opinion is more correct?

The situation of Russian science journalism is very special. On the one hand, positive steps have been taken which show that an attempt has been made to create new standards of journalism as it plays a major role in the European community. And on the other hand, the reality is rather incompatible with the country which used to be

M. Claessens (ed.), Communicating European Research 2005, 191–194.

known for many years as the home of intelligent and well educated persons, with fine examples of science.

There are probably three periods in Russian science journalism.

The first one was at the time of the former Soviet Union where science journalism was mostly a part of official propaganda. And at the same time, science journalism was a refuge for those in the profession who did not want to write about a "happy communist future". Science was respected in society and the circulation of science popular media was extremely high. A science section could be found in every newspaper.

The second period belongs to the beginning of the Nineties. The collapse of the economy left no opportunity for the development of science. As a result, there was no interest for science journalism either. Science desks in newspapers, TV and radio were closed. Media became independent and had to earn money and editors started to look just for hot stories about politics, crime, sex, show business – anything which could help to sell media production. Nobody cared about science. Media gave readers stories about paranormal and extrasensory events, astrology, crime, sex and so on. It was a period when there were tough standards of pseudoscience in mass media. It was also a period where many science journalists and writers changed their field of interest.

The third period began at the end of the Nineties.

First of all, it was a result of the efforts of scientists and science journalists who never gave up, even in the difficult aforesaid period. They continued their efforts to give the public the most interesting results of research and the most important is that science itself got some more support in the country. The development of i-technologies was a great help. One could read best European, American, and Australian information via Internet. Nowadays very successful examples of Russian Internet sites could be seen, with interesting and useful science popularized information. The first site was created by science popularisation agency, Informnauka (*www.informnauka.ru*). A very successful site belongs to the Foundation of Professor Dmitry Zimin "Dinastia" (*www.elementy.ru*). There are some other popularized resources: *www.spacenews.ru*, *www.inauka.ru*, *www.poisk.ru*, *www.pereplet.ru*, *www.astronet.ru*, *www.scientific.ru* and others.

Since the end of the Nineties, there were special science pages in national papers such as Rossiyskaya gazeta, Izvestia, Nezavisimaya gazeta, Literaturnaya gazeta, Moscowskaya pravda, Parlamentskaya gazeta, Trud and so on. These pages differ greatly and are sometimes not of very good quality. But – anyway – newspapers now no doubt pay attention to science. The circulation of science popularisation magazines is still low and some of the magazines collapsed (e.g. Lomonosov, Paradox), but there were some new ones (What's New in Science and Techniques, several magazines in conjunction with well known foreign ones such as Scientific American and others).

The situation is worse on radio and TV: there are a few programmes on radio ("Radio Rossii", "Majk", "Ekho Moskvi"), but one can hardly find any science popularisation programme on national TV channels and if so, never in prime time. For example, Professor Sergey Kapitza's programme "Clear – unbelievable" is only broadcast after midnight. This same programme was no longer on the TV for about a decade during the Nineties and is now back on again!

Do these examples mean that the current situation of science journalism is agreeable? No.

Here are some more opinions about science journalism in Russia. Last year, the Institute for psychology belonging to the Russian Academy of Science asked scientists to answer the question: "Do you like the way Russian media cover science?" According to the report of the Deputy Director of that Institute, Dr. Andrey Jurevich, only 3% of respondents said that their impressions were positive. 33% of respondents wrote: "Negative". 37% answered that they have no specific opinion. 27% expressed their opinion and here are some examples: "Journalists cover science with stereotypes, without any competence", "Journalists often lie, they do not understand what they are writing about", "Those who write well about us belong to American, Japan and other foreign media, but Russians write only badly about us", "Russian journalists prefer to write about the poor salary of scientists, they forget the science itself", "Russian media do not cover science at all".

It is obvious that scientists are mostly negative about journalists.

The same result was obtained during an Internet conference organised 2 years ago with the help of the Konrad Adenauer Foundation, the Institute for Social Sciences belonging to the Russian Academy of Science and the agency Informnauka. This week-long discussion was published as a book entitled "Russian Science and Media". Looking through the pages, one can see that scientists blame journalists and journalists express their negative emotions after contacts with scientists. Here is a quote by a researcher from the Sternberg State Astronomical Institute Dr. Sergey Popov: "Scientists and media have different goals and different mentalities. Scientists follow their ideas and their research. Journalists always stay under the pressure of hot stories. Scientists and journalists have nothing in common". One is reminded of the words of Rudyard Kipling who wrote one day:

East is East and West is West
And never the twain shall meet.

Should the same be said of science and journalism?

It is true that there are not enough journalists who can write about science and here one can recall the situation of the beginning of the Nineties when numerous professional journalists left science and began to cover other fields which were better paid. It is true to say that editors do not give much space to science yet. But it is also true that Russian scientists have no impulse to speak with journalists about their research because they do not see what will be changed after any publication. Scientists do not see what they will get after collaboration with media. They cannot say like American professor Neal Nisbeth who wrote once: "If you'll speak with journalists they will tell lies for sure, but if you'll ignore journalists you'll lose money". Unfortunately the Russian state of economy is not yet that positive and science and scientists cannot see any direct impact after association with media.

But anyway scientists need journalists even in Russia because media can change the public's attitude towards science.

And journalists need scientists, because only scientists will help them to write a good story. So there must be communication.

Russian science journalists try to change the situation. One line being followed is to offer training to journalists. For example, there are study trips organised by the European Union of Science Journalists' Association (EUSJA) and competitions organised inside the country for science journalism. In 2005 the fourth Competition was organised by the International Science Center, the British Council, the agency Informhauka and the Russian Association of Science Writers and Journalists Intellect. The Russian Academy of Science and the Ministry of Education and Science were involved this time in the organisation - a very positive sign for the future.

Another sort of training is science cafés. There are 3 different styles of science cafés in Russia:

1. for the public (the British Council in Russia organised several meetings in different Russian cities during November 2005 with the main topic "AIDs"); -
2. for journalists: they meet in cafés and discuss different subjects (e.g. "What is pseudo-science and how one should it be avoided in writing); these meetings are organised by the radio science journalists (radio "Svoboda") Alexander Kostinsky; –
3. for journalists and scientists: they meet in cafés with a strong agenda, journalists listen to scientists and then scientists listen to what journalists have to say. There are different topics – the manner in which journalists write about the environment, medicine; events in physic's year and Aids...These meetings are organised by the International Science Center and agency Informnauka and, they are very useful for science communication.

And now a new project is being designed which could help bring together Russian scientists, the public and the media. This is a project of the Russian Science Festival (Lomonosov Festival). The Russian Association for Science Writers and Journalists agreed with the Moscow State University that this biggest educational and research centre will be a main stage for the science festival, and Moscow State University is already planning the First Russian Science Festival for the coming October. This should be based on co-operation between EUSJA and EUSCEA. This is just the beginning and the project is still in need of improvement and funds. It is rather difficult to find funds for a major science popularisation project in Russia. It is difficult as well to organise international study trips for science journalists. There are many hopes that the new project with science festival has a future.

CHAPTER 34

HOW TO REACH THE BUSINESS MEDIA?

GILL JOY[1], MARTA RIBELE[2], SEAN DUKE[3], MICHAELA STIPSITS[4], LUISA MINOLI[5]

[1] *IST Results Service (http://istresults.cordis.europa.eu/)*
[2] *'Dienas Bizness' (www.db.lv), currently 'Diena' (www.diena.lv)*
[3] *'Science Spin' (www.sciencespin.ie)*
[4] *'Business People' (www.businesspeople.at)*
[5] *'Innovare' (www.rivistainnovare.com)*[1]

Abstract: There are two justifications for increasing efforts in communicating science to a wider audience: firstly, there is a responsibility to inform taxpayers how their money is being spent on publicly funded research (and perhaps gain their support for new and increased research initiatives), while secondly there is a role for communication in getting new technologies to the marketplace. Given Europe's challenge in transforming research results into commercial success, it was this second role that was the main focus of a forum discussion organised by the Community Research and Innovation Information Service CORDIS (*http://cordis.europa.eu/*) to address the issues of science-business communication. The specialist business press aims to reach managers and executives within companies that might well be interested in taking up research results from European projects for use in commercial products and services. Business editors from around Europe provided insights into their work and advice on how best to reach the business media

Keywords: Science-business communication, Research results, Business press, Public-private partnerships

1. INTRODUCTION

Science and business are still seen as two worlds apart, but they both need each other and business journalists are among those who can help bridge the gap. The panel discussion involving business editors from all around Europe sharing their

[1] For bios and presentations, visit CORDIS at: *http://cordis.europa.eu/guidance/past-events2005.htm#-communication*

M. Claessens (ed.), Communicating European Research 2005, 195–199.

experience in covering science for a business audience started from the understanding that:

> Readers of the business press, who are mainly managers and executives, are not looking for technologies, but for business solutions, and understanding the needs of the business community is THE KEY for being able to reach their specialised press.

2. SCIENCE NEWS IN A BUSINESS DAILY

MARTA RIBELE, news editor of 'Dienas Bizness', Latvia's daily business newspaper and part of the Scandinavian Bonnier Group, began by addressing some of the reasons why scientific news does not have a higher profile in the press. A lack of well presented information reaching journalists was the chief cause, according to her, at least partly due to the culture of science itself, which is rarely business oriented. Scientists do not necessarily pass on their information to the media, or in the case of smaller countries like Latvia, they may be more interested in sending information to foreign, international media.

The business media is, by its nature, mainly interested in stories based on people or money, and this should be borne in mind when deciding how to present research information in an interesting way. Scientists should consider angles likely to be useful or of interest to readers who may seek to 'mirror' themselves in the article.

She justified this by saying that initially the editors of Dienas Bizness considered that news about EU Framework Programmes was not for them. Since there is money involved, however, and the issue of how Latvia uses EU funding is high on the public agenda, interest in stories about how this money is used has driven increased media attention in results from EU-funded projects. She went further to recommend that organisations that co-ordinate national participation in the programmes provide access to databases of projects and results for interested journalists.

3. PRESENTING SCIENCE IN A BUSINESS CONTEXT

MICHAELA STIPSITS, managing editor of 'Business People', a magazine supplement to Austrian business newspaper 'Wirtschaftsblatt', emphasised the importance of flexibility and creativity ('thinking outside the box') in spotting opportunities to translate science stories into business stories:

- Topical issues, such as avian flu, can increase business interest in research areas such as vaccines;
- Features of popular scientists;
- End-of-year reviews or previews can offer the opportunity to contribute technology forecasts and present new developments;

- Finding an angle for the business reader, like how is the story related to money, do the research results help to meet legal requirements?
- Her presentation was backed up by practical examples of scientific stories in the "Business People" magazine.

4. DO'S AND DON'TS IN WORKING WITH THE PRESS

Based on his experience in both science and business publications, Sean Duke, who is the editor of "Science Spin" magazine, contributed a presentation that stressed the following points:

- Research makes a good business story when the science is new and has direct relevance for industrial development;
- As with any news story – the wider the impact on people, the better the story;
- Don't use jargon, don't try to control what is written, don't take too long to reply and don't disturb the journalist either with repeated calls or at deadline time;
- Imagine things from the editor's point of view, not your own:
- Is this science new, will it make companies more competitive?
- Is it important to my readers, how many business people will it affect?
- Can I tell a good story; is there any human drama?

'Think of editors as intelligent, half alert children that get bored easily!' was his concluding message.

Editors give below some practical advice on how to deal with journalists directly:

- First of all, send an introduction by e-mail;
- Then follow this up with a phone call that should be polite but concise and to the point (but they agreed that this should not be done at deadline time);
- If successful in grabbing an editor's attention, the interview or phone interview could be followed up with a one-page overview that clarifies and summarises key points that should not be missed;
- Maintain good relations with the journalist and be trustworthy;
- Provide contact details such as phone number and web address which allows the journalist to research details in his/her own time.

A proactive approach to getting on to a journalist's radar could even involve sending some information twice. However, one should be particularly confident that it is of real interest. Suggestions for story angles and features or supplements are also usually welcome.

Editors recommended putting oneself in the position of the journalist, especially as deadlines approach. Thoroughly researching the publications that one is targeting is an important first step in deciding how to sell a story. Again, flexibility is key, and if a magazine is not currently interested in a biotechnology breakthrough, the possibility that it has already received or is looking for venture finance may be more relevant for an article on venture capital, for example.

Journalistic contacts can also be cultivated by showing willingness to answer journalists' calls and questions when looking for a scientific viewpoint on another story.

5. PARTNERING WITH THE SPECIALISED PRESS

LUISA MINOLI, representing 'Innovare', the magazine of the Italian Confederation of SMEs, addressed some of the specific questions associated with reaching smaller firms, a vital audience for the transformation of research into innovation. She emphasised the diversity of the business media, pointing out that many specialist magazines that reach SME managers are interested in working closely with researchers in order to present their readers with detailed information on new technologies' applications and benefits.

Researchers are advised to focus their story on 'business solutions' and best practice when trying to bridge the scientific to business worlds. They should explain the benefits, such as improved quality or reduced waste, and include key information such as time to market, potential savings, etc. Existing case studies, based on covering EU-funded projects show the way in which business magazines can collaborate with research teams in order to develop well-presented articles.

Editors suggested building partnerships with business associations and information services (such as CORDIS and IST Results) since they have newsletters that may be less dependent on daily deadlines.

6. TIPS FROM THE ROUNDTABLE DISCUSSIONS

Just one of the practical tips that arose from the authors' experience is that researchers or coordinators should subscribe to business PR wire services. These will already have a distribution list for the most relevant business publications, it was noted, so most of the work has already been done in getting a project's press release to the right journalist. IST Results articles are sent to press wire services where they have been taken up by thousands of journalists, thus helping to raise IST projects' international visibility.

Today, some people think that web technology has made paper publications obsolete – researchers could put their information online themselves and the ever more sophisticated search engines available would ensure that it is found by the correct target audience.

However, the authors stress that journalists and editors base their decisions on what topics will interest their readership, and are ever mindful of the need to attract readers' attention enough to persuade them to sit down and read the article. While this can make publication more challenging, the greater impact it can have is one area where participants felt traditional paper publications still have an advantage over Internet media.

Against this background, it is worth pointing out that while CORDIS is a 100% web service, its print digest, CORDIS *focus* magazine, often attracts new potential users of the service and has a proven record of being favoured by project co-ordinators wishing to raise awareness of their achievements and/or international events. Internet and print publications are thus not contradictory, but complementary channels to use. This complementarity can be used both ways: the IST Results service has become a good channel for reaching the print press, as for example the "Financial Times" newspaper.

The editors pointed out that the impact of the article, and even the likelihood of its publication can be improved by supplying or offering to supply photographs. According to some survey statistics, photographs could be even more important than text. They help in grabbing the attention of readers and persuading them to spend time reading an article during the 20 minutes or so they spend browsing through the paper.

Media competition has led to small editorial teams with few journalists and therefore a shortage of time. There is no time to go for information indirectly so information must be presented in a practical way to stand a better chance of being used as a story.

Access to company figures could sometimes be an issue due to a lack of trust of journalists. Magazines have an interest in protecting their reputation by getting stories right – they have no desire to create scandal out of nothing and therefore sometimes send advance copies of articles to check facts (though not with the aim of receiving approval).

Looking at longer-term approaches, there is a need for closer partnerships between companies, universities, business associations and information channels. Such partnerships can make it easier to bridge the different approaches characteristic of the scientific and business worlds.

The practice of 'embedding' journalists, allowing them to spend an extended visit in laboratories and among researchers at work, is an experimental approach that holds a great potential for future improvements in science's place in the media.

ACKNOWLEDGEMENTS

This is an edited version of an article that first appeared in CORDIS news (http://cordis.europa.eu/news/) in November 2005.

CHAPTER 35

THE SAME OLD FUTURE

CORMAC SHERIDAN

Science Journalist, 61 Aughrim Street, Dublin 7, Ireland, Tel +353-1-8689306,
Email csheridan@eircom.net

Abstract: A greater historical awareness of previous media depictions of science and technology could help to improve the contemporary practice of science journalism and enrich its contribution to debates on bioethical issues. However, co-opting science journalism in the pursuit of public policy goals appears to be a more pressing priority

Keywords: Jacques Loeb, Craig Venter, Synthetic biology, Bioethics, History of science, Science policy, Media

Contemporary biological research during any given era seems to reanimate the same controversies that have previously been played out during earlier periods. Ethical concerns, which are rooted in philosophical, cultural and religious beliefs about the sanctity of human life, appear largely constant, certainly when set against the pace of research.

1. BEFORE VENTER

In 2002, the news that the Institute for Biological Energy Alternatives (IBEA, now part of the J. Craig Venter Institute) in Rockville, Maryland, USA, had received funding from the U.S. Department of Energy to construct a synthetic genome encoding the 'minimal gene set' necessary to support life elicited dramatic headlines:

- 'Scientists to synthesize new life form' UPI, – 21 November 2002
- 'Scientists hope to create new form of life' – Reuters, 20 November 2002
- 'How one scientist has made it his mission to break the ultimate taboo: to create life itself' – The Independent, 22 November 2002

M. Claessens (ed.), Communicating European Research 2005, 201–203.

A century earlier, the publication by Jacques Loeb (1859–1924), of a paper [1] on the chemical induction of parthenogenesis in sea urchins provoked a remarkably similar response:

- 'Creation of life – startling discovery by Professor Loeb. Lower animals reproduced by chemical means.
- Process may apply to humans. Immaculate Conception explained. Wonderful Experiments Conducted at Woods Hole' – Boston Herald 1899
- 'Science nears the secret of life: Professor Jacques Loeb develops young sea urchin by chemical treatment. Discovery that reproduction by this mean is possible is a long step towards realizing the dream of biologists: to create life in a test tube' – Chicago Tribune 1899

German-born Loeb can, in several respects, be seen as the Craig Venter of his day. His pursuit of an engineering-based approach to biology, which had as an explicit goal the physico-chemical manipulation of life, exerted a profound influence on the development of the life sciences in the 20th century. Venter's exploits in genome sequencing and genome synthesis are rooted in that branch of the discipline which Loeb shaped and developed.

Like Venter, Loeb also enjoyed a high media profile and a degree of public notoriety. However, apart from a scholarly biography published in 1987 [2] and a handful of references in historical articles, the name of Jacques Loeb seems to have been largely forgotten among the current generation of scientists and among the journalists who report on their work. Yet, while the content of their respective experimental programmes obviously differs, it is evident that the popular responses that Loeb and Venter evoked bear striking similarities.

Loeb's work is, of course, no longer the subject of ethical scrutiny or societal concern. While parthenogenesis was later artificially induced in several other animals, including species of frog and salamander, the process did not apply to humans (and it certainly did not shed any light on the Catholic Church's doctrine of the Immaculate Conception, which had been established as dogma in 1854). Research – at the J. Craig Venter Institute and elsewhere - on synthetic biology remains a significant cause for concern. Contemporary societies are also still grappling with the dilemmas engendered by research on human embryonic stem cells and on somatic cell nuclear transfer.

A greater awareness of the wider intellectual and historical contexts surrounding contemporary molecular biology and cell biology could help to inform the very necessary democratic debates that will, hopefully, influence their responses. Journalists - be they specialists in science or in other relevant areas – have a potentially useful role to play in these debates.

2. CO-OPTING JOURNALISTS

Whether, however, there is any space within contemporary science journalism for such an historical dimension is an open question. Other priorities are receiving more attention at present. It is evident that science journalism – and science journalists – are being co-opted by governments and policy makers to support and promote science in order to assist in the attainment of science policy goals, among them, an intensification in the performance and in the commercialisation of scientific research; and the promotion of science to young people in order to ensure a future supply of science graduates in Europe.

Whatever the merits or demerits of these goals are, the notion that science journalists should in some way participate in the task of recruiting the next generation of scientists suggests a fundamental misunderstanding of their role. The same agenda would never be applied to the specialist journalists and correspondents that cover politics, economics, legal affairs or education, for example. If the regular pattern of media coverage of science and related activities results in greater numbers of students opting to undertake science in school or at university, well and good. But that can only be a side effect, however welcome it may be.

REFERENCES

"On the Nature of the Process of Fertilisation and the Artificial Production of Normal Larvae (Plutei) from the Unfertilised Eggs of the Sea Urchin," American Journal of Physiology 3(1899):135–138

Pauly PJ (1987) 'Controlling Life: Jacques Loeb and the Engineering Ideal in Biology', OUP

CHAPTER 36

EUROPE IN SPACE – TAKING OFF WITHOUT THE PUBLIC?

DIRK H. LORENZEN

Science Writer, Board Member of German Science Journalists' Association (WPK), Wildgansstraße 32 c, 22145 Hamburg, Germany, Tel +49-40-67941029, Email dirk-wpk@gmx.de

Abstract: Several examples show that there is a lack of proper science communication in Europe. Europe is doing great in astronomy and space flight, but it has a hard time to communicate that. It seems that American scientists are more enthusiastic about communicating their research than most of their European colleagues. The big question is then: Why? Is it a cultural difference? Is it political pressure to get the next project funded by communicating current activities? Is there more staff in public affairs departments in the US than in Europe?

Keywords: Astronomy, Communication, Media

January, 4th 2004: the Mars Rover Spirit has just landed on the surface of Mars. It transmitted the first colour images from the Gusev crater, presented in near real-time in a big press event, headed by NASA administrator Sean O'Keefe.

January, 19th 2004, 16:30 CET: the first image of Mars taken by the ESA spacecraft Mars Express is published: it pops up in the internet without any special presentation event, but with a quite angry public who are fed up with waiting for Europe's first Mars image any longer. The spacecraft had been in orbit around the red planet for several weeks already. An official event with high level representatives from ESA and politics follows four days later.

November 2005: the Cassini spacecraft has been in orbit around Saturn for almost 18 months now. Each day a beautiful image from Saturn is published. More than 500 Cassini images are online. After almost two years of operation, a mere 50 images of ESA's Mars Express are online.

The Hubble Space Telescope (HST) – interestingly enough, it is a joint NASA/ESA undertaking, with a 15 per cent European share – publishes marvellous pictures almost on a weekly basis. Europe's organisation for astronomy, the

M. Claessens (ed.), Communicating European Research 2005, 205–208.

European Southern Observatory (ESO), operates the Very Large Telescope (VLT) in Chile, considered the world's most powerful astronomical instrument. In 2004, ESO published just five "Hubble class" VLT pictures.

These are just a few random examples of a lack of proper science communication in Europe. Europe is doing great in astronomy and space flight, but it has a hard time to communicate that. The bottom line is quite obvious: American scientists are more enthusiastic about communicating their research than most of their European colleagues.

The big question is: Why? Is it a cultural difference? Is it political pressure to get the next project funded by communicating current activities? Is there more staff in public affairs departments in the US than in Europe? Does the average US scientist prefer sharing his or her impressions with the public more than most of his or her European counterparts? Is it an issue of public perception of science – with US citizens being more science minded than most European citizens, many of whom consider Mozart, Voltaire and Michelangelo as culture but not Kepler, Galileo and Newton? It might be a mixture of them all.

However, it is the scientists' job to leave their ivory towers and to get their news around. If science does not matter to the public, the public will lose interest in it. In the long term this might end in a decrease of funding. While this might pose a severe problem for important but unappealing science in the future, it could be a great opportunity for astronomy and space flight activities. Both are easy to communicate. Astronomy and space flight fire imagination, they have the appeal of extreme conditions and extreme numbers, they are producing beautiful images almost everybody will get fascinated by. It seems that not many European scientists and engineers in those fields have realised that – or that they suffer from constraints like political structures that hamper public relations work.

In a stroke of genius, the team of HST has devoted about one percent of its observing time to the Hubble Heritage Project. Its only purpose is to produce beautiful images. They are presented as artwork, not necessarily in the proper scientific orientation etc but in the most appealing way to the public. Does this project mean that the HST team is wasting time? Would it be "better" to use this time for "real" science (although in most cases those beautiful images are of scientific value anyway)? Today, HST is the most famous scientific instrument in the history of astronomy. To a great extent this is due to almost perfect communication activities (done by a huge team of course, numbering dozens in certain periods).

And Europe? Of course, the four VLT telescopes are used mostly in spectroscopic mode – not taking images of the objects in the sky but splitting the light into different colours. Spectroscopy is more meaningful scientifically. If we assume the four telescopes can be used on average 80 percent of a year, we end up with well over 1000 VLT observing nights. What would it cost to devote ten nights per year to imaging just for PR (public relations) purposes? ESO might end up with a few publications less in peer-reviewed papers. So what? Is the 115th article in a scientific journal really more important than a good picture making the front page of newspapers throughout Europe? There are many astronomical pictures taken by HST that are familiar to almost everybody. Many people will not know what the image shows and that HST has taken it. But many pictures are almost scientific icons or symbols of our time. Where is ESO? There are only very few well-known VLT images – and many people refer to them as HST pictures. HST will be history within the next five years. Astronomy will lose its picture machine. Is Europe willing to take over? A few dedicated nights per year could cause an avalanche of public outreach data, public perception and reputation.

Communication is visual, mostly. But communication is done *with* images; it is done *by* human beings, i.e. PR officers and scientists. What is the payoff for the scientists? Is it of any use to invest time into public outreach activities? Well, it is great in the spirit of science, but as a matter of fact, scientists do not benefit from their own outreach work immediately. However, in the long run, it may increase his or her visibility and public awareness of this particular research area. Is this something a scientist should care about? At least young scientists should...Applying for an academic position, everyone is asked to bring a publication list. However, if an applicant acknowledges to have a few peer-reviewed publications less because he or she had spent on average two hours per week talking to journalists, had given public talks and had written a popular book, nobody will really care. It does not count in the academic world. What is wrong with our science system?

The Principal Investigators of instruments flying on European satellites have a very strong position. In most cases, they virtually own the data. They decide whether specific data are used for public outreach or not. They decide what to show the public early in the mission (after some years all the data are public, mostly). How is it possible that a great deal of money is spent on space and astronomy, but in the end, the scientists responsible for the projects are the only ones to decide how much the public should get to see of it? Is scientific equipment the private toy of the relevant researcher? The press officers of various institutions suffer from this ignorant behaviour by some scientists.

Why is public relations work not an integral part of any project financed by the European taxpayer? Since FP6, public relations activities are mandatory according to all contracts granted by the EU. But it does not seem that anybody really cares about it. Nor is a certain share of the budget allocated for outreach purposes. Does anyone check whether and how the communication of a specific project is done? Of course, it is quite hard (next to impossible) to judge whether PR activities have been well done or not. The success depends on many factors the players have no control over. Is it not possible to ask any applicant for examples of his or her outreach activities and press coverage in former projects? We need a shift in attitude. Scientists need to realise that communication is necessary and that it has its benefits.

Astronomy and space flight have much more appeal to many people than many other parts of science and engineering. Space events have the potential to attract kids to follow a professional career in science and technology. There is a whole Apollo generation of scientists and engineers (not only astronomers and space flight engineers!). In the era of the Lisbon process we can not afford to waste this potential any longer. We need action taken now. Excellent science and efficient event-driven communication are no contradictions. A major discovery will turn out to be a hot shot for communication too. Good pictures may be excellent tools for communications maybe even without any scientific meaning.

Europe is doing great in space. But Europe's achievements are hardly communicated. Scientists, public outreach officers and journalists need a sound basis to live and work together. They need a rigid discussion on how to get along with each other and on what their respective tasks are. Europe is doing great in space – but we all need a completely new attitude towards the communication of science.

CHAPTER 37

POPULATION EXPOSURE TO AIR POLLUTANTS IN EUROPE (PEOPLE)

How clean is the air in European Cities?

P. PÈREZ BALLESTA, R. A. FIELD AND E. DE SAEGER

European Commission, DG JRC, Institute for Environment and Sustainability, TP-441, 21020, Ispra (VA), Italy, Tel +39-0332-785322, Email pascual.ballesta@jrc.it

Abstract: The PEOPLE (Population Exposure to Air Pollutants in Europe) project is presented. With the support of the media and thousands of citizens, at six European cities (Brussels, Lisbon, Bucharest, Ljubljana, Dublin, Madrid), samples of personal, indoor and outdoor air pollution concentrations were collected. Air quality in cities is controlled by the relationship between emissions of and dispersion of pollutants. Transportation and environmental tobacco smoke were important factors for exposure to the air pollutant benzene. While collecting scientific data the project also helped to raise awareness of the public to air pollution in general and more specifically to the importance of personal behaviour for exposure to air pollutants

Keywords: Air Quality, Air Pollution Emission Sources, Exposure to Air Pollutant Benzene, Public Participation, Environmental Monitoring Design

1. PEOPLE – HOW CLEAN IS THE AIR IN EUROPEAN CITIES?

The PEOPLE (Population Exposure to Air Pollutants in Europe) project assessed indoor, outdoor and personal exposure levels of air pollutants in European cities.

The main aims of the PEOPLE project were:

- to support national, city scale, air pollution monitoring network design;
- to evaluate the importance of indoor and outdoor emission sources for personal exposure to the air pollutant benzene;
- to compare air quality in various European capitals and identify similarities and differences;

M. Claessens (ed.), Communicating European Research 2005, 209–215.

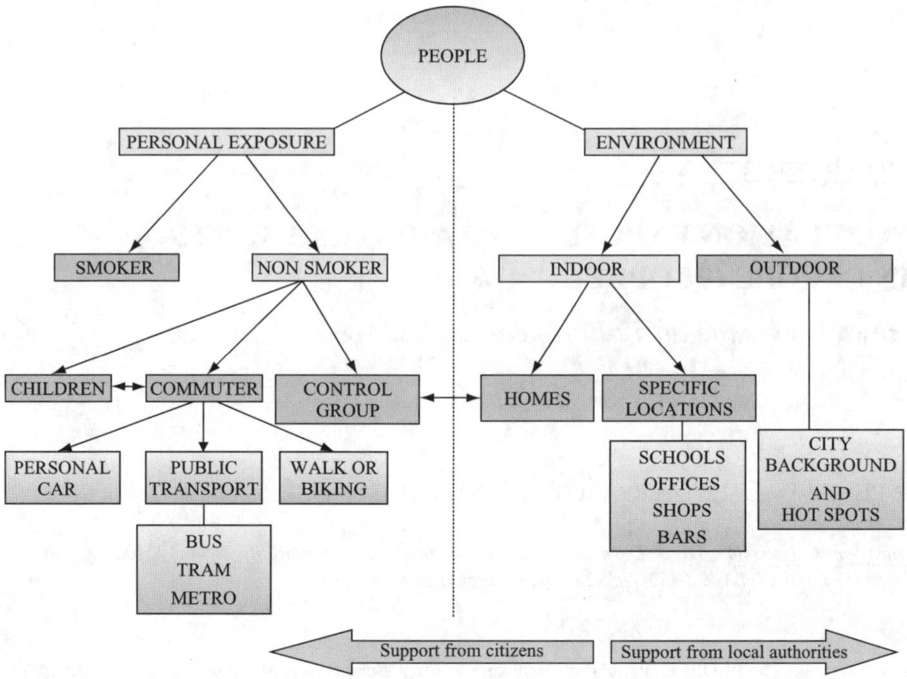

Figure 1. Schematic illustrating the project design

- to raise the awareness of citizens with regard to air quality in general, and in particular to the impact of lifestyle.

Six European cities were studied, namely: Brussels, Bucharest, Dublin, Lisbon, Ljubljana and Madrid. In each city, diffusive samplers were used to monitor personal exposure and environmental pollution levels of benzene. The focus of the project for personal monitoring was to determine exposure levels for normal city workers. Personal monitoring was performed for commuters with typical behaviour profiles. The definition of the sampling groups was linked to the two main sources of pollution: transportation and smoking.

Benzene was measured as it is the first carcinogen to be regulated by EU air quality directives (2000/69/EC). Benzene is associated with the risk of the development of leukaemia. As a primary pollutant emitted by traffic, benzene is also a good marker for other pollutants generated by traffic, such as carbon monoxide, nitrogen oxides and volatile organic compounds. Smoking is also an important source of benzene strongly affecting personal exposure to this pollutant. It should be noted that benzene is, however, one of many toxic compounds emitted by smoking, such as nicotine, tar and fine particles. In cities, benzene is mainly generated by automotive traffic. It is present in petrol and is emitted from motor vehicles through evaporation and from the exhaust system as either a combustion product or as unburned fuel.

Citizens from participating cities were invited through the media (television, radio and newsprint) to fill out a screening questionnaire to determine if they were suitable as a volunteer for the study. The questionnaire was designed to determine how long and by what means people travel to work. Citizens were then selected to represent differing groups of transport, for example those that travel using personal car, or those that travel using public transport, or those that travel on foot or by bike. A small group of about thirty citizens represented each commuting group.

Up to 200 people were selected to carry monitors in the particular city under investigation for a period of 12 hours. These people commuted from home, usually in the suburbs, to work in offices, usually in the city centre. A wide variety of different professions were represented including teachers, journalists, nurses, electricians, librarians and bankers amongst many others. The monitors were collected, analysed and compared to both a control group that consisted of people who work or stay at home and a group of smokers. The results could be linked to lifestyle as each participant recorded all their activities and the places they visited. By completing this diary, that also indicates transport modes and presence of tobacco smoke, a complete profile of the day with a 5 minute time resolution, was possible.

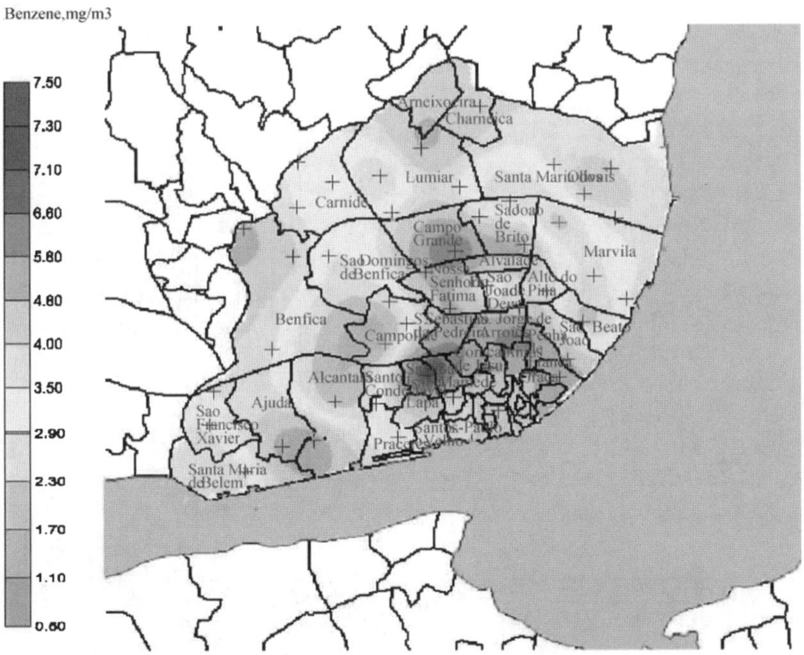

Figure 2. Background levels of benzene in Lisbon 22 October 2003

The exposure of the people in each city to the air pollutant benzene was also compared to the pollution of the environment, both outdoors in the city and indoors in the most frequently visited places: houses, offices, schools, restaurants and bars, as well as inside some transport modes like buses, cars and taxis. The measurements of the indoor environments provided information on the importance of smoking. Extensive spatial monitoring of air quality provides information across a wide range of outdoor locations. Such data enables pollution contour mapping that illustrates the background air quality or pollution profile across the city. These contour maps are complimentary to the information that is gained from monitoring stations that operate to check compliance to air quality directives.

The main results of the PEOPLE project are:

- The snapshot design of the project successfully provided good scientific data while including citizens in environmental research.
- Contour mapping complimented network monitoring data and in Bucharest showed suitable areas for the placement of air quality stations.
- Personal exposure to the air pollutant benzene was linked to lifestyle, in particular transport and tobacco smoke. Volunteers that smoked had the highest level of

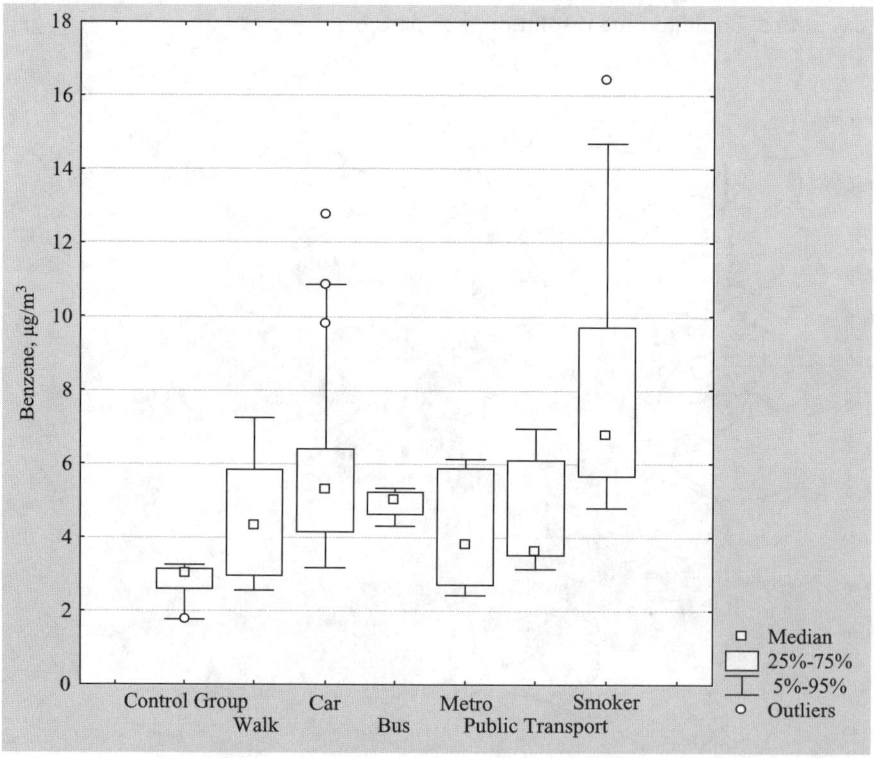

Figure 3. Personal exposure levels of benzene in Lisbon 22 October 2003

exposure. Of the commuters, car users had the highest exposure due to closer proximity to air mixed with emissions from car exhausts.

- The general level of exposure was related to the background air quality of the city. There was a relationship between the outdoor air quality and exposure of the population. The ratio between population exposure (non-smoking commuters) and background air quality was approximately two to one. This reflects the importance of movement through "hot spots" and travelling at busy times when pollution levels are high.
- The highest levels of environmental pollution were reported in enclosed spaces related to transport or indoor emission sources. Cars travel through the position of transport emissions whenever moving, so high pollution levels are anticipated. The highest level of indoor pollution level was associated with smoking. In Dublin, where a smoking ban is in force, environments like pubs and restaurants are now equivalent to the outside air.
- The lowest level of air pollution in the six cities studied was measured in Dublin and the highest was measured in Bucharest. The relatively good air quality in Dublin is due to enhanced pollution dispersion due to high winds whereas the poor air quality in Bucharest is due to high levels of emissions from transport. In this city the fuel contained much higher levels than EU member states that have legislation in place to control the content of benzene in petrol to less than 1%. Also the exhaust emissions from cars are higher as the vehicles are generally older and do not comply to the most recent EU standards.
- The project has enabled the development of an estimation approach to population exposure through the development of "exposure factors". This approach may be useful to contrast the relative importance of different activities and places across European cities.
- The information from the project acts as a platform for consideration of more complex particulate pollutants that are or will be controlled by EU air quality Directives such as PM_{10}, $PM_{2.5}$, heavy metals and polycyclic aromatic hydrocarbons. The control of these pollutants is recognised as important for the protection of human health.

Since air quality directives sets standards for acceptable levels of air pollution in outdoor air to protect human health, it is important to understand the relationship between actual population exposure and background air quality. Comparison of data from personal and

Table 1. Comparison of PEOPLE Cities 2002 to 2004

	Estimated annual median spatial city background benzene level ($\mu g/m^3$)	Annual Exposure level ($\mu g/m^3$) for non-smoker commuting population
Brussels (2002)	3.4	5.1
Lisbon (2002)	4.1	6.1
Bucharest (2003)	8.6	12.9
Ljubljana (2003)	3.7	5.5
Madrid (2003)	3.4	5.1
Dublin (2004)	1.9	2.9

environmental monitoring, both indoors and outdoors, has helped to better define the important factors for population exposure to air pollution.

The involvement of the media and the public ensured the project was not only useful for scientists but also for a general audience, including the participants and city wide community. The interest of the results was greatly enhanced by working with the a city. Pollution levels, like the news, are more interesting when they have local relevance. People are more engaged when information is from their own city, in particular when they help to collect it. The inclusion of volunteer citizens and the rapid presentation of results in the media increased the visibility and impact of scientific data that would otherwise be consigned to academic papers and policy support reports. The success of the project to meet the awareness aim has been determined to a large extent by the level of local interest created. The importance of the media cannot be underestimated. With the media as a partner outreach of environmental policy into the population can be improved. It is important that the public understand why environmental data is collected. Communication with an appropriate level of detail depends upon the target audience. It is clear that more than simply providing information is needed. The public may not know why some pollution issues are important. As such the value added step that makes people aware of and understand issues is of importance. This requires clear and simple messages that are inclusive of the whole population. If a message becomes too complex more becomes less and the audience can be lost.

While scientific research establishments often have close links with universites, schools are often distant. The inclusion of school children into the project, both as volunteers and through an extension of the project into classrooms in Lisbon, was a novel approach. This enabled a step forward from information supply towards assisting environmental education. In Dublin a half an hour environmental program called Ecoeye on the main national TV channel RTE, was devoted to the PEOPLE project and associated issues. In Lisbon classroom debates extended the project to become an educational tool. The media and classroom coverage in Dublin and Lisbon were the pinnacle of the outreach and awareness raising achieved by the PEOPLE project. Through this project the European Commission directly engaged the citizens that it serves.

REFERENCES

Ballesta PP et al (2006) Population exposure to benzene: One day cross-sections in six European cities, Atmospheric Environment 40(2006):3355–3366

Field RA et al (2005) Population exposure to Air pollutants in Europe (PEOPLE) Methodological strategy and basic results, EUR 21810 EN, EU report

ACKNOWLEDGEMENTS

We would like to thank the following people and organisations that worked on this project: A. Baeza Caracena, R. Connolly and N. Cao of European Commission, DG JRC, Institute for Environment and Sustainability; P. De Mulder and A. Meurens of Brussels Institute for the Environment, Belgium; A. Cabral of Institute for the

Consumer, Presidency of the Council of Ministers, Portugal; L. Nogueira of Regional Administration of Environment and Land Use Planning for Lisbon and Tagus Valley, Ministry of Cities, Land Use Planning and Environment, Portugal; P. Ferraz de Abreu of Research Center on Information Technology and Participatory Democracy, Portugal; P. Otorepec and M. Gregoric of Institute of Public Health of the Republic of Slovenia; E. M. Niciu of Institute of Public Health Bucharest, Ministry of Health, Romania; R. Fernandez-Patier and D. Galán Madruga of Institute of Salud Carlos III. Ministry for the Health of the Consumer, Spain; P. Goodman of Dublin Institute of Technology, Ireland; and M. Fitzpatrick of Dublin City Council, Ireland.

CHAPTER 38

COMMUNICATING EU FOOD AND HEALTH RESEARCH

TORGER BOERRESEN[1], PI HÖGBERG[2], GEORGE CHRYSSOCHOIDIS[3], FILIP CNUDDE[4], TERESA BELCHER[5] AND JÖRG OEHLENSCHLÄGER[6]

[1]SEAFOODplus, Danish Institute for Fisheries Research, Department of Seafood Research, DTU building 221, DK-2800 Lyngby, Denmark, Tel +45-45-252577, Email tb@difres.dk
[2]FP6COMMNET and CASCADE, Department of Biosciences and Nutrition, Karolinska Institutet, S-14157 Huddinge, Sweden, Tel +46-8-6089237, Email c.johansson.cascade@biosci.ki.se, present address: Swedish National Institute of Public Health, Akademigatan 3, 831 40 Östersund, Sweden, Tel + 46-63-199660, Email pi.hogberg@fhi.se
[3]TRACE/EuroFIR, Agricultural University of Athens, Laboratory of Agribusiness Management, Iera Odos, 75 Votanikos, GR-11855 Athens, Greece, Tel +30-210-5294766, Email gc@aua.gr
[4]SAFE FOODs, Wageningen University, Hollandseweg 1, NL-6706 KN Wageningen, The Netherlands, Tel +31-317-482406, Email Filip.Cnudde@wur.nl
[5]Med-Vet-Net, Society for Applied Microbiology, Bedford Heights, Brickhill Drive, Bedford, MK41 7PH, UK, Tel +44-1234-271020, Email teresa@sfam.org.uk
[6]SEAFOODplus, Federal Research Centre for Nutrition and Food, Palmaille 9, D-22767 Hamburg, Germany, Tel +49-40-38905151, Email joerg.oehlenschlaeger@bfel.de

Abstract: FP6COMMNET is an informal network of communication managers in 15 research consortia which receive funds from the European Union (EU)'s Sixth Framework Programme. About 3000 scientists are fully or partly engaged in the consortia forming the FP6COMMNET at present. The members of FP6COMMNET address common issues related to the design and implementation of communication activities within European Union's funded projects in the health and food sector

Keywords: Food and health research, Communicating research, European Union's Framework Programmes

1. INTRODUCTION

The authors are members of FP6COMMNET, an informal network of communication managers in research consortia which receive funds from the European Union (EU)'s Sixth Framework Programme (FP6). Members of FP6COMMNET are the following

M. Claessens (ed.), Communicating European Research 2005, 217–222.

consortia: SAFE FOODs, GA2LEN, DIOGENES, LIPGENE, SEAFOODplus, CASCADE, NoMiracle, EuroPrevall, EADGENE, NeuroPrions, NuGO, MED-VET-NET, TRACE, EICOSANOX, and EuroFIR. Communicating EU food and health research is important for non-governmental organisations (NGOs), authorities, and other policy-makers that are stakeholders in Community research, European Commission officials, science journalists, especially those with special interest in the biosciences, ongoing FP6 consortia, newly FP6 research consortia, planned FP7 research consortia and other groups interested in the management of science-society issues and/or communication of the biosciences.

2. THE ROLE OF SCIENCE COMMUNICATORS

Why do we need science communicators? Scientists have always known how to do research and how to educate graduate students; however, they do not often have the skills to communicate successfully beyond the scientific community. Moreover, they do not have the time; often do not see the need to communicate to a broader public and some are unwilling to do so due to lack of experience, confidence, incentives etc.

One reason for this is that public communication is a specialist skill. As communication managers form a link between scientists and stakeholders/society, they must be multi-skilled and must have profound knowledge of the scientific activities in their respective projects.

Communication managers must also be able to meet the European Commission's requirements for effective dissemination of research results and achieve successful integration with stakeholders. They should also bridge the gap between the scientists of his/her own consortium and the public.

Then comes the question: Who do we communicate with? "Internal communication" means setting up a communication infrastructure, preparing and distributing regular newsletters, creating a private website, building online discussions for organised internal meetings, holding active continuous communications with partner institutes by means of networking being present at presentations and attending conferences. "External communication" includes media relations, television, radio, and contact with print journalists and web journalists.

By using media communications, the public and non-scientific stakeholders can ultimately be reached and press releases and/or press conferences are an ideal tool for the dissemination of research results.

Networking is another tool for making contacts, establishing relationships, forming new collaborations and sharing experiences – particularly within FP6 networks of excellence (NoEs) and integrated projects (IPs).

The marketing of the consortia themselves and of results obtained within the consortia by writing articles for external magazines and journals, by editing press releases, brochures, public magazines and newsletters, as well as having exhibition stands and producing giveaways with a link to the project is important too.

For communication managers, training in science communication is essential. This can be done e.g. by attending short courses in science communication, workshops in crisis communication or lectures on the subject of cultural differences in communicating. Furthermore, the approval of a joint communication plan is highly recommendable.

Science communicators must co-operate closely with their administration in reporting to the European Commission and project members, and when preparing the annual report. Science communicators have to be involved also in the editing and translation of reports and press releases, in budget responsibility, in financial accounting, in staff management, in the organisation of meetings and all kinds of events

In conclusion, communication managers have a vital role in EU research consortia. They need multi-skills and have to work according to current best practices. This is all of utmost importance because good communication of their projects leads to greater understanding of science by stakeholders and the general public.

As the overall aim of our projects within the food quality and safety area is to improve consumers' health and wellbeing, successful communication will make these requirements a reality. Lastly, better science renders European industry more competitive by means of improved co-operation, reduced duplication of research and more funding opportunities.

3. HOW TO REACH YOUR STAKEHOLDERS?

There are two important questions which communication managers must ask:

- Who is your target audience and
- Who are the stakeholders of your project?

In order to be able to answer these questions, communication managers first have to ascertain the project goal. Is it to develop a specific framework or model or is it to develop a specific test or technology or to generate knowledge in a precisely and well defined domain (health, food, allergy), or is it to create a network to exchange knowledge? Or is the project aiming at preventing allergy, enhancing food safety, or solving nutritional problems or is it about training or education? After they have defined the answers to the first questions, and before selecting dissemination tools, communication managers must identify the stakeholders involved.

The aforesaid stakeholders can be, for example, consumers, patients (or their organisations), a defined or undefined research community, health professionals, policy and decision-makers, media professionals, communicators, food or other industries, NGOs, food safety authorities or intergovernmental organisations. Once they have identified the stakeholders, communication managers have to define their message to them and to target their audience.

Success stories of the manner in which to involve stakeholders are for example: active participation of industry partners in EU projects and with groups of patients, having stakeholders as members of project advisory boards or as participants in meetings of the consortia and in conferences, and also the organisation of specially designed stakeholder consultation events.

In conclusion, the challenges for a communication manager are the following: to hold good balance of internal versus external communication, involve all project partners equally in communication, ensure dissemination across all EU countries, have an overview and understanding of all ongoing research activities within the project, check regular flow of information and develop strategies to maximize output in a limited period of time.

4. IMPLEMENTATION OF RESEARCH OUTCOMES IN/BY 'SOCIETY'

In this section we are using the two following definitions: *research* is a "careful, systematic study to establish facts and principles to advance existent knowledge, and to develop technology" and *innovation* indicates a series of operations and processes whereby such facts, principles, technology or knowledge (created, disseminated and distributed) are utilised.

The results of innovation are made evident in competitive products, industrial processes and services.

Against this background, the diffusion of innovations can be defined as the process by which the innovation is adopted into a social system.

The position of industry regarding the creation and use of the result of research and innovation is that industry forms an important part of the social and knowledge system and consequently must fully participate in the creation of research-led innovations per se. Industry is the main mediator of the link between research and consumers.

What is the manner in which one can speed up the adoption of the results of research and innovation by industry and later on by society? Important factors are:

1. any innovation must contain a relative benefit, which means it must be better than the idea it supersedes,
2. good compatibility, which means consistency with past experiences and the perceived and real needs of receivers,
3. a certain and limited degree of complexity, meaning that it is not too difficult to understand and use,
4. a trial run which makes it possible to experiment over a limited time, and finally
5. observability, by which it is made visible to others.

The adoption of innovation by industry is influenced by different issues like firm technology strategies and existing investments, the respective management of product development processes, the new product concept development, the quality of links with the research system, the entrepreneurship and the quality of internal organisational design.

The adoption of innovation by organisations relies on developments in the environment of the different industrial sectors and the stakeholders' influence. But to a great extent, it also depends on internal company practices for new product development and the adoption of research-led knowledge and innovation.

In conclusion, research-led innovations need to possess specific characteristics for faster adoption. Industry has a mediating role regarding implementation of research in, or by, the wider 'society' and further understanding of what is going on within the industry and firms in the different sectors to speed up the adoption of the outcome of research. Innovation must be based on research.

5. COMMUNICATING WITHIN EU-FUNDED PROJECTS

IPs and NoEs within the food quality and safety areas of FP6 receive funds up to EUR 17 million per project. However, total budgets are much higher than the EU contribution, as many partners have contracts granting only 50% of their total efforts. About 3000 scientists are fully or partly engaged in the 15 consortia forming the FP6COMMNET at present. Most of these were granted as a result of the first wave of proposals under FP6. More projects have been granted since, and new projects will be adopted as they come through the negotiation stage and start operating. Furthermore participation is encouraged in the European Technology Platform 'Food for Life', which is mainly industry driven.

Within the EU-funded projects, the following issues are addressed by FP6COMMNET:

- How does one communicate with potential new partners in countries not represented in the consortia when these countries have different languages (e.g. new Member States)? Strategies that can be used were suggested like, for example local meetings using the respective language or communication at a simple and generally understandable level.
- Based on their experience, FP6COMMNET members agree that a separate budget for communication in FP7 is absolutely necessary to disseminate information from the scientific consortia to scientists outside of it and to people in the street and that this money cannot be taken out of the limited management budget. This separate budget is also necessary to train scientists in media skills or to hire professionals to do the job.
- The extent to which industry is the mediator between researchers and society and a necessary link in the implementation of results can be discussed. Different approaches exist in relation to this issue. In order to obtain the effective transfer of technology, large multinational companies are sometimes ideal partners. But there are also other bodies which SMEs (small and medium-sized companies) should target. How can researchers best reach the industry? There seems to be a consensus for establishing specific platforms or for identifying target groups in industry. Furthermore, it is considered a "must" to also include transfer of money into development and transfer of knowledge into industry. If industry is not involved in a project and stands outside of it, it can be involved and pulled in via platforms. Ideally, however, the industry should be directly involved in a project as a partner. In addition, industrial associations must be approached and invitations for conferences should preferably be sent to them as they can distribute the material further to their members.
- Presentations must be balanced to embrace different target groups and also groups who are critical of the project. The knowledge should be made understandable for everyone in order to form a collective conscience and to re-open closed minds.

- Can basic research be communicated? The authors answer yes, but the research should be in a form that is understandable to all levels of the public and is clearly thought through beforehand to avoid creating a situation of panic. FP6COMMNET members can provide interesting examples.
- According to the European Commission, there might be better guidance under FP7 on how to use the budget for the purpose of communication. The recommendation for new or ongoing projects in FP6 is, however, to put communication issues in separate work packages and to make proper budget allocations for them.

CHAPTER 39

COMMUNICATING ENVIRONMENTAL RESEARCH

WILLY DE BACKER[1], ANDREW TERRY[2], ANDRÁS DEMETER[3], BARBARA DEMENEIX[4], PIERRE COËRS[5], JACQUES DE SELLIERS[6]

[1]*EurActiv.com, International Press Center, B-1041 Brussels, Tel +32-2-2265814, Email willy.debacker@euractiv.com*
[2]*World Conservation Union (IUCN), Boulevard Louis Schmidt 64, B-1040 Brussels, Tel +32-2-7328299, Email andrew.terry@iucn.org*
[3]*Nature and Bio-Diversity Unit, Directorate-General Environment, European Commission, 200 rue de la Loi, 1049 Brussels, Belgium, Tel +32-2-2963245, Email andras.demeter@ec.europa.eu*
[4]*Muséum National d'Histoire Naturelle, UMR CNRS 5166 Evolution des Règulations Endocriniennes and Dept Régulations, Développement et Diversité Moléculaire, Rue Cuvier 7, F-75231 Paris, +33-1-40793607 Email demeneix@mnhn.fr*
[5]*SOLVAY Corporate Management for Research and Technology, Health and Safety Environment, Rue du Prince Albert 33, B-1050 Bruxelles, Email pierre.coers@solvay.com*
[6]*GreenFacts asbl/vzw, Rue des palais 44, B-1030 Brussels, Tel +32-2-2113425, Email jacques.deselliers@greenfacts.org*

Abstract: GreenFacts organised a Forum Session at CER 2005 entitled "Communicating Environmental Research". The discussion acknowledged the difficulties environmental journalists encounter in conveying scientific views, which are often uncertain. News stories were found to be the key to environmental reporting, whereas background information or multi-faceted issues such as biodiversity were seen as following a steeper path up to the news. Participants and audience alike reflected on the implications of scientifically-based environmental reporting for decision-making in democratic societies

Keywords: Environmental research, Risk, Science communication, Journalism, News, Health, Democracy

The first presentation was made by Mr. András Demeter, from the European Commission, who identified some strengths and weaknesses in the relationship between scientists, policy-makers, journalists, and the public, particularly in the field of biodiversity. Biodiversity loss is an example of key environmental issues that are often poorly communicated in mainstream media. Predictions are usually based

M. Claessens (ed.), Communicating European Research 2005, 223–225.

on statistical models and indicators, which are difficult to digest for decision-makers and difficult to summarise for the wider public.

Then, Mr. Andrew Terry, from the World Conservation Union (IUCN), explained how his experience in communicating with policy-makers reflected similar problems to those discussed by Mr. Demeter: it seems much easier to communicate climate change than biodiversity losses; in general, Mr. Terry believes that environmental processes, as opposed to "events", are inherently difficult to communicate. He also stated that policymakers are partially responsible for communicating environmental issues to the citizens, and advised research managers to take into account communication needs from the beginning of their research plans, and not as an "end-of-pipe" exercise.

Mr. Pierre Coërs, from the Solvay Group, focused on the communication of environmental risks. Mr. Coërs defined the term "risk" as "a probability of danger" which needs to be properly managed and becomes the subject of a political decision, which should ideally be based on science and democratic principles. In communicating environmental research, the expectations of the public and the "expert" answers may be quite the opposite: for instance, the public demands absolute certainty in terms of safety margins, whereas risk-free solutions do not normally exist. Furthermore, Mr. Coërs believes that no factual explanation, however brilliant it might be, will succeed in convincing an anxious public. Efforts to avoid anxiety must be made beforehand. The National Museum of Natural History in Paris, represented by Professor Barbara Demeneix, mentioned that the REACH draft Directive does not seem to take into account the high risks of more than 50,000 chemicals to which Europeans are currently exposed, and of which only 5,000 are tested for side effects on health. She emphasized the problem of making scientific approaches which address chemical contamination in the environment more interesting to journalists. Professor Demeneix also stressed that the difficulties in communicating science to the general public started because communication channels between scientists and journalists were not clear in the first place.

Mr. Jacques de Selliers, from GreenFacts, focused on the need for clear and reliable information on the environment tailored for journalists. In Mr. de Selliers' view, a scientific consensus document produced by large panels of experts is a great starting point for improving communication on any given environmental research. He described how GreenFacts conveys the scientific consensus available in those documents by collaborating, for instance, with the World Health Organization (WHO), the Food and Agricultural Organization (FAO), and the European Commission.

Mr. de Backer, from EurActiv, said that, in his view, the key issue for journalists was the newsworthiness of the issues at stake: journalists need precise stories and have little time to dig into the background. Furthermore, they have to choose among competing stories.

The first question of the discussion dealt in depth with the need of background explanations for readers. Another person from the audience wondered if there is an ideal "environmental journalist"; Mr. de Selliers acknowledged that it would be difficult to find a one-size-fits-all description but, in his opinion, most journalists have an important responsibility due to the high impact they produce on public perceptions and policy decisions.

One Norwegian attendee stated that environmental journalists were not necessarily a condition *sine qua non* of fluent communication between the public and policy makers or scientists, and mentioned some pilot schemes which had taken place at local government level in Norway which put citizens directly in contact with decision-makers on agricultural and environmental conflicts.

Finally, Mr. dc Backer thanked the participants and the audience and expressed again the wish that the communication of environmental research be better tailored to the needs of journalists in a responsible and democratic policy-making environment.

CHAPTER 40

TALKING NANO – WHAT MAKES NANOTECHNOLOGY SPECIAL

"While technology shapes the future, it is people who shape technology, and decide what it can and should be used for" (Kofi Annan)

RICHARD HAYHURST[1], WOLFGANG M. HECKL[2], GUGLIELMO MAGLIO[3], VOLKER TÜRK[4], DAVID BENNETT[5]

[1]*4Bio and Hayhurst Media, Hayhurst Media, 7 Kilmorey Gardens, St Margarets, London, TW1 1PU, Tel +44–208–8912177, Email richard@hayhurstmedia.com*
[2]*Ludwig-Maximilians-Universität and Deutsches Museum Munich, Museumsinsel 1, 80538 München, Germany, Tel +49–89–2179313, Email heckl@deutsches-museum.de*
[3]*Fondazione IDIS-Città della Scienza, via Coroglio 104, 80124 Naples, Italy, Tel +39–081–5702158, Email maglio@cittadellascienza.it*
[4]*Wuppertal Institute, Doeppersberg 19, 42103 Wuppertal, Germany, Tel +49–202–2492152, Email: volker.tuerk@wupperinst.org*
[5]*Delft University of Technology, Julianalaan 67, 2628 BC Delft, The Netherlands, Tel +31–15–2127800, Email david.bennett@efbpublic.org*

Abstract: Nanotechnology is entering a phase where the involvement of the public is becoming crucial in terms of public acceptance, market potential and political governance. This chapter reviews the issues involved in communicating about nanotechnology with the broader public: its novelty, ethical, legal and social aspects, and comparison with debates surrounding other new technologies such as genetic modification. It also covers ongoing activities, best practices and further research needs. The chapter also presents three projects currently funded by the European Commission in this area: NanoDialogue, Nanologue and Nanobio-RAISE

Keywords: Nanosciences, Nanotechnology, Media, Science Communication, Nanomanipulation, Self-assembly, Nanobiotechnology, Science museum, Public acceptance and debate, Ethical, Legal, Social Aspects, GMO debate

M. Claessens (ed.), Communicating European Research 2005, 227–232.

1. TALKING NANO – MEDIA ISSUES AND TRENDS FOR NANOTECHNOLOGY

The challenge today is to create a positive image of nanotechnology amongst the public through the media based on understanding, excitement, trust, as well as personal and societal benefits. But this means no longer just dealing with science specialists. Editors, staff, and freelancers are active in news areas such as social affairs, economics, industrial, environment, politics, health and lifestyle, nutrition, religion and show business. Excitement, scandal, personalities are what every journalist wants in a story. The only answer is to accept this and actively work to maintain a balance.

How is nanotech doing? On the surface, it is well covered. A Google search brings 47 million hits for the key word "nanotechnology", 1.6 million hits for "nanomaterials" and 0.25 million for "nanomedicine". But they are not yet reaching highly educated audiences. A review of high quality journals show a low coverage with only a few lead articles: "Nature" – 719, multi-journal, old leads; BBC – 4 mentions, Radio, 4 leads; "Der Spiegel" – 17 in a year, no leads; "Sciences et Avenir" – 7 in a year, no leads. Nor does nanotechnology reach the general public. An analysis of Lexis-Nexis shows 716 mentions of nanotechnology in last 6 months in worldwide newspapers compared with 560 about biofuels, for example, during a week at the beginning of November 2005.

The conclusions to be drawn are:

- Nanotechnology needs excitement – just like internet has in the past
- Nanotechnology needs personalities – young, dynamic scientists
- Nanotechnology needs to save the planet *and* fulfil dreams
- Nanotechnologists need to understand the media – and to use it.

2. TALKING NANO – WHAT MAKES NANOTECHNOLOGY SPECIAL

At the dawn of the new century a powerful technology opens the doors to a new world in which we can visualise and manufacture on the nanoscale. The scanning tunnelling microscope, invented by Gerd Binnig and Heinrich Rohrer and honoured with the Nobel Prize, is key to this nanoworld. Experiments of nanomanipulation like the vertical transposition of a C_{60} guest molecule or writing at the atomic scale are only first steps in the direction of imitating nature's self-organisation and self-assembly principles.

Besides long term wishes and visions like:

- the production of new advanced materials based on quantum effects rather than classical physical phenomena
- the fabrication of tiny machines to replace Nature's machines (Nanobionics)
- the curing of widespread diseases or
- solving the water shortage problem. . . .

We also have to be aware of and discuss about the risks connected with this new technology.

More information on the activities of Prof Heckl at LMU Munich can be found on www.nano-science.de.

3. SCIENCE CENTRES AND PUBLIC DEBATES IN NANOTECHNOLOGY: THE NANODIALOGUE PROJECT

Do Science Centres involve visitors in debates about science and ethical/social/legal issues? Often we have been hearing about the need to establish a dialogue between science and society. If people do not understand the role of science and technology then they will not be able to support the right policies for the future. But even if this need is acknowledged by scientists, research institution, media and stakeholders in general, there are not many specific places where the dialogue can occur and involve citizens as well. Fondazione IDIS – Città della Scienza, aims at creating a new science community that requires the science providers to join a dialogue with the consumers of science in the form of the visitors and general public.

The NanoDialogue project, funded by the European Commission and coordinated by Fondazione IDIS, aims at 'Enhancing a dialogue on Nanotechnologies and Nanosciences in society at European level' by providing information and raise awareness among general public on nanotechnologies and nanosciences. It aims also at implementing a dialogue between the research community, civil society and citizens, and tests high quality communication tools and participatory methodologies. Above all, it develops a common exhibition module, displayed at the same time (March-September 2006) in science centres in eight European countries (Technopolis – Belgium, AHHAA – Estonia, CCSTI – France, Deutsches Museum – Germany, Città della Scienza – Italy, Ciencia Viva – Portugal, Barcelona Cientific Park – Spain, Universeum – Sweden). The partnership is complemented by ECSITE, the European Network of Science Centres and Museums, and the Centre for Studies on Democracy of the University of Westminster in the UK.

Some of the problems raised by nanotechnology research turn out difficult for people to understand and discuss, since they are felt like "not concrete" issues. Science fiction can help to overcome this obstacle, since it deal with people's most common fears and emotions about new technologies, plus using a language more accessible to the large public breaking down conceptual barriers. That is why the exhibition starts with a science fiction module, where scenarios coming from the most famous novelists are discussed and analysed. In each section, together with the main information, there are comments on the ethical, legal and social aspect of the topics in order to let visitors confront their ideas and opinion with those expressed by scientists, politicians, philosophers and science fiction writers. The scientific part of the exhibition displays models of nano-particles together with real applications. A central table, representing a space for debate and discussion, gives the opportunity for people to "play" with a hands-on exhibit and real nano-objects; at the same time the space can be used for science demonstrations and small debates, while multimedia programs give visitors the opportunity to write their comments and to read what previous visitors have said.

During the display of the exhibition, different events will be performed in all countries, in order to enrich the exhibition module, but also to give visitors the opportunity to be involved more actively with forums and workshops in the debate. Feedback will be collected also through a structured questionnaire that will be analysed by the Centre for Studies on Democracy (UK), in order to provide recommendations and suggestions to the European Commission during the final event taking place in February 2007.

More information about the project can be found on www.nanodialogue.org.

4. NANO-CHANCES, NANO-RISKS: THE NANOLOGUE PROJECT

The field of nanotechnologies has clearly attracted widespread attention and funding in recent years. The unique properties of nanotechnological applications suggest potential to solve some of the worlds most pressing challenges, but they come with uncertainties and risks as do all new technologies. Taking advantage of technological progress and preventing adverse side-effects and societal backlashes requires analysis, evaluation and guidance to ensure the technology is developed in ways that benefit the economy, wider society and the planet. Aiming at contributing to this goal, the Nanologue project has been set up.

Nanologue brings together researchers, businesses and civil society representatives from across Europe to support the dialogue on the societal implications of nanotechnologies. Funded by the European Commission, the project is driven by the need to understand ethical, legal and social aspects (ELSA), i.e. benefits and potential impacts, of nanotechnologies – and communicate this understanding by raising awareness and providing information to societal actors. Nanologue is led by the Wuppertal Institute (Germany) and conducted in cooperation with EMPA (Switzerland), Forum for the Future (UK) and Triple Innova (Germany). The project, which started in Spring 2005 and will last until mid 2006, comprises three main steps:

1. A mapping study on recent publications and developments regarding selected nanotechnology applications and ethical, legal and social aspects (ELSA) to lay a common ground for the subsequent discussions.
2. Moderated dialogue sessions for an inclusive and neutral platform for information and opinion exchange and discussion. Interviews with experts to substantiate findings and opinions.
3. Scenarios that translate the insights gained for easy communication on the potential implications of these emerging technologies.

Results of the project are disseminated by a variety of means, ranging from media workshops, a website to a project pamphlet and conference attendances. An outcome of specific interest for businesses and researchers will be the "Nanotechnologies Opportunities and Threats Checker" (NOTC). Planned to be a free of charge online tool, NOTC is intended to help "translating" the rather abstract terms of ELSA, sustainability, societal concerns etc. into something more tangible for those being involved in the development and marketing or NT-based applications.

In addition to businesses and researchers, NOTC may well facilitate educational purposes. NOTC offers students for example in the natural sciences or business administration fields to gain insights of and awareness relevant for ELSA and broaden their perspective beyond their area of study. The multidisciplinary nature of nanotechnology along with the number of societal issues and questions concerned make nanotechnology-based applications and NOTC interesting study objects for students, university researchers etc.

The latest project results, the studies and reports and further information can be found at www.nanologue.net.

5. NANO-BOOM, NANO-HYPE AND NANO-TALKING – THE PROJECT NANOBIO-RAISE

Nanotechnologies are expected to have a profound impact on societies and economies around the globe. At the same time, public awareness of nanotechnology is very low. This has led many people to expect that nanotechnology, and especially its nanobio-applications, will be the next major public, NGO, media and political issue after GM food and agriculture and following on others such as with nuclear energy.

As part of the effort to anticipate such a back-lash, the European Commission recently established the Nanobio-RAISE project. Nanobio-RAISE brings together the key players in the field including nanobiotechnology researchers, ethicists, communication specialists, policy makers, company representatives and non-governmental organisations to:

- horizon-scan for the scientific and commercial developments likely to cause public and political concern,
- clarify the ethical issues and public concerns involved or as they arise, and recommend and carry out strategies for public communication to address the emerging questions.
- take on board the experiences and lessons learned from the European GM debate of the last decade and apply them with this project to the nanobiotechnology discussions.

One of the lessons, hard-learned, from the GM debate is that the vast majority of European people are not really interested in science, do not understand it and do not want to unless they have a personal need to. Otherwise their interest in science and technology is as spectacle, entertainment or controversy. The answer would seem to be to concentrate on a range of public communication and engagement activities which experience shows do have significant genuine effect and which are possible within the given circumstances. Scientists see the need to communicate and engage with the public. Some do it very well but many are hesitant for reasons of priority of research, publication and career, lack of training and confidence, and lack of tangible rewards. But they have a responsibility to explain their science and its importance for the general good, as well as for their own good as scientists, and because they are paid to do their science via public funding.

The Nanobio-RAISE project takes its lead from these experiences. Expert horizon-scanning workshops will be organised to anticipate the likely social and ethical issues in the various areas of nanobiotechnology. Specific issues relating to human enhancement will be discussed in depth by an expert group. Public opinion focus group discussions will be held in different parts of Europe to assess regional variations in public perceptions and opinions. The knowledge gained through these activities will provide the input for the project's communications programme. Nanobiotechnology scientists will be encouraged and trained in the many different ways of communicating and engaging with the public with advanced public communication and ethics courses. Three series of briefing papers will discuss in-depth the issues involved for the general public, teachers, media, etc., for government officials and politicians, and for scientists. A series of ethics lectures during scientific conferences will increase awareness of these issues with the scientific community. Public relations will make all the project's results known to politicians, media and the public.

More information about the project can be found on www.nanobio-raise.org.

CHAPTER 41

COMMUNICATE INTERNATIONALLY – WITH PARTNERS FROM THE NEW INDEPENDENT STATES (NIS)[1]

RICHARD BURGER[1], TATIANA RUNGE[2], LIUBOV STRELNIKOVA[3], KAMILA MAGZIEVA[4], VLADIMIR KOMLEV[5]

[1]*Science & Technology Counsellor, Delegation of the European Commission to Russia, Kadashevskaya naberezhnaya 14/1, 119017 Moscow, Russia, Tel +7-495-7212000, Email richard.burger@ec.europa.eu*
[2]*ININ Programme Manager, INTAS, 58/8 Avenue des Arts, 1000 Brussels, Belgium, Tel +32-2-5490111, Email info@intas.be*
[3]*Chief Editor, InformNauka Agency, Lefortovsky per.8, 105005 Moscow, Russia, Tel +7-495-2619793, www.informnauka.ru*
[4]*National Information Point (NIP) Co-ordinator, InExCB-Kz, 28 Shevchenko Street, 480021 Almaty, Kazakhstan, Tel. +7-32-72726360, www.nip.kz*
[5]*Scientist, Russian Academy of Sciences, Ozernaya 48, 119361 Moscow, Russia, info@intas.be*

Abstract: The purpose of this chapter is to foster exchange of communication on best practices between the NIS, European scientific communities and media organisations. Particular attention is given to the skills of young scientists to communicate with the public, as demonstrated by Vladimir Komlev, winner of the INTAS Young Scientist Writing Competition 2004, who was invited by several publishing houses to write reviews based on his winning article. The experience also provided him with the opportunity to develop a network of European partners

Keywords: International co-operation, New Independent States

1. BACKGROUND

In 2003, with the support and assistance of INTAS (International Association for the promotion of cooperation with scientists from the New Independent States of the former Soviet Union), a network of National Contact and Information Points for

[1] The forum has been organized by ***INTAS*** (www.intas.be), *the* International Association for the promotion of co-operation with scientists from the New Independent States of the former Soviet Union.

M. Claessens (ed.), Communicating European Research 2005, 233–235.

the EU's Sixth Framework Programme 2002–2006 (FP6) was set up by NIS partner countries, with the aim of raising awareness about the European Research Area (ERA) and about European funding opportunities amongst NIS scientists and the media. Building on the successful launch of the Information Points, the communication focus of the network now includes highlighting to EU Member States the valuable scientific expertise available in the NIS for potential partnerships.

2. HOW CAN WE COMMUNICATE WITH OUR NIS PARTNERS?

The moderator, Richard Burger, opened the session by stressing the importance of sharing experience and building on best practice for future successful international scientific and technological (S&T) co-operation. Prime examples of information exchange through a network of contact points are the'China-EU S&T Cooperation Promotion Office' (CECO) in Beijing and the INTAS FP6 NIS Information Network (ININ), which had been established to promote the involvement of their scientific communities in ERA and their participation in FP6. The ININ Programme Manager, Tatiana Runge explained further how the established network of contact points in the NIS served as an information multiplier, using brokerage events to encourage networking and participation in new calls for proposals. Activities and events targeted at the scientific community were also communicated through training sessions, regular e-mail updates and articles published in the leading Russian-language weekly scientific journal "Poisk".

3. 'IF IT DOESN'T EXIST IN THE MEDIA, THEN DOES IT EXIST AT ALL?'

Still on the theme of communicating science through the media, Chief Editor Liubov Strelnikova illustrated the power of the Internet through InformNauka (see also page 191), the only agency in Russia to specialize in science news over the past six years. With weekly news updates in the fields of science, technology and medicine, plus a short internet version, the agency links up with foundations, embassies, scientists, journalists and the media to communicate science-related events and research progress speedily and effectively. Other innovative features of the agency include a regular Science Café, which provides a lively and entertaining forum to discuss different aspects of science, and a "Science for Society" competition with categories for the best science photograph, and best articles on popular science and Russian innovations.

4. A KAZAKH SCIENTIST'S PERSPECTIVE

Kamila Magzieva, InExCB-Kz (Kazakhstan) looked at communicating science from the scientists' viewpoint, and gave an overview of the benefits of Kazakhstan's partnership with INTAS and the EU Framework Programmes. In particular she explained that although the budget for research & development in Kazakhstan was limited to priority fields, participation in INTAS and FP INCO calls had given Kazakh scientists an opportunity to learn how to initiate international research projects, to network with partner countries, protect their property rights and raise the profile of their research by attending international scientific events. Access to the INTAS e-library facility provided up-to-the-minute information on the latest scientific developments worldwide but, as Dr. Magzieva indicated, perhaps the most valuable tool in opening up the EU RTD platform to the NIS was the establishment of an FP6 National Contact Point in Kazakhstan, resulting in the creation of a website (www.nip.kz) to provide regular electronic news updates on FP6-related thematic information days, training and seminars.

5. MOBILITY FOR YOUNG SCIENTISTS FROM NIS

Winner of the INTAS Young Scientist Writing Competition in 2004 Vladimir Komlev homed in on the need to nurture the new generation of scientists to strengthen international R&D and to build a global Research Area. Current sources of support for young scientists, such as the INTAS Young Scientist Fellowship scheme and Marie Curie grants, provided an opportunity for talented young scientists to pursue cutting-edge research in their chosen field, and to communicate with leading European institutions. However, Mr. Komlev stressed that more funding was needed to support the mobility of young scientists, enabling them to develop their own teams through extended networks and disseminate the results of their research to a wider audience.

6. HOW CAN WE COMMUNICATE BETTER WITH NIS PARTNERS?

Some of the current issues include the possibility of providing training for NIS scientists to improve communication with the media, the importance of mobility for young scientists, dissemination of research results at international conferences and difficulties encountered in translating research results. In conclusion, the moderator summarised the key elements necessary for successful international S&T cooperation, such as relevant up-to-date databases, consistent and reliable information updates, targeted brokerage events, visible and sustainable information initiatives with long-term impact and, last but not least, the active support of authorities in partner countries.

CHAPTER 42

HOW TO COMMUNICATE AN INTERDISCIPLINARY PROJECT?

Communication within and on interdisciplinary projects inside the scientific world and to the public

CHRISTIANE WEHLE

EU-Bureau of the Federal Ministry of Education and Research, Co-ordinator NEST-IDEA, Königswinterer Str. 522, 53227 Bonn, Germany, Tel + 49-228-447646, Email christiane.wehle@dlr.de

Abstract: New opportunities and global problems often involve an interdisciplinary approach when they are tackled. In this context there are a number of barriers to interdisciplinarity that need to be solved. Amongst these barriers several have been identified in the area of research communication. Integration of disciplines can only be accomplished through initiation of communication among and between the respective scientific communities and practitioners. The forum "How to communicate an interdisciplinary project" provided an opportunity to explore different aspects of communication within and on interdisciplinary projects inside the scientific world and to the public. The objectives were to learn about the experiences of others and share own success stories. Issues that were discussed concerned management and communication structures within interdisciplinary projects including barriers and advices specific to them. In addition, good practice for communication to the scientific community and to society in general as well as communication tools were explored

Keywords: Interdisciplinary communication, Management of interdisciplinary research, Communication tool for interdisciplinary research

1. INTRODUCTION AND OBJECTIVES

NEST-IDEA is a European Union funded action which aims to give support to the scientific community and the European Commission regarding new and emerging science and technology (NEST) initiatives and to furthermore promote the dialogue with the research community on emerging scientific and technological developments.

M. Claessens (ed.), Communicating European Research 2005, 237–241.

This article addresses issues on management and communication structures within interdisciplinary projects including barriers and advices specific to them. In addition, good practices for communication to the scientific community and to society in general as well as communication tools are presented.

A forum organised at CER 2005 provided an opportunity to explore different aspects on communication within and on interdisciplinary projects inside the scientific world and to the public. The latter involved presentations of successful cases of communication among project partners and instruments of disseminating results by experts and coordinators of interdisciplinary projects.

The article is divided into three main sections. The first section presents basic rules for communication inside interdisciplinary projects as well as to the scientific world and to the public. The second section of the article deals with communication structures within an interdisciplinary project and in this context the example of the CREEN project are presented. The third section concerns communication on an interdisciplinary project outside of the scientific world, towards stakeholders and citizens, and presents the experience of the CONTROL CANCER STEM project.

2. BASIC RULES FOR COMMUNICATION WITHIN AND ON PROJECTS INSIDE AND OUTSIDE THE SCIENTIFIC WORLD

Kathrin Stratmann from the EU-Bureau of the Federal Ministry of Education and Research (Germany) presented the two different levels of communication relevant to the topic:

1. Communication *within* the project (with partners, work package leaders, coordinator), with the goal to ensure the flow of information between the project participants.
2. Communication with the *outside world* (scientific world, laypeople), with the target to present the results.

2.1. Communication within the Project

Within a project, the coordinator has to communicate on different levels with different partners (e.g. the project partners or the European Commission). For this, it is important to keep in mind that a consortium always incorporates a variety of partners (different personalities, countries, cultures, institutions, levels of experience).

In interdisciplinary projects additional problems may occur concerning the different approaches, the personal scientific background and the different (discipline-specific) languages of the project partners.

It is recommended to establish a common language at the beginning of the project. Constant communication throughout the project is essential for its success.

2.2 Communication to the Scientific World and to the Public

One example for communicating an interdisciplinary project to the scientific world is to publish the results in scientific journals while communication with laypeople or the interested public often is provided via internet, TV or talking to target groups (for example at schools).

The problem which occurs in interdisciplinary projects is that different disciplines require different approaches when communicating to different target groups. Knowing the target group and addressing its specific needs (for example in using keywords) is one of the basic requirements.

A good strategy for communicating with the outside world is to provide a concerted approach through a website, publications or presentations, to plan ahead and to choose the approach with keeping the targeted audience in mind.

3. COMMUNICATION STRUCTURES IN INTERDISCIPLINARY PROJECTS

Daniel Bieber, from the Institute for Social Research and Socio Economics (ISO, Germany) reported on the outcomes and recommendations of the expert group established by NEST-IDEA on the evaluation, management and general background of interdisciplinary projects.

The group suggested four relevant topics that need to be dealt with if with a view to the funding as well as the management of inter-or transdisciplinary projects:

1. Evaluation (evaluators are often experts in just one discipline),
2. Structures (main recommendation is to overcome path dependencies at universities),
3. Management (the coordinator of interdisciplinary projects needs also social skills)
4. Communication (there is a higher need for communication outside and inside of interdisciplinary projects):
    - Communication outside the project: There is a higher need to legitimate the interdisciplinary research question or problem and the new methodology towards the different disciplines and faculties.
    - Communication inside the project: All participants of a project have to learn "new languages" and a common understanding has to be established within the project partners (for example of the research problem, of the methods or of the products).

The expert group recommends finding common ground in areas of interdisciplinary research which needs special forms of organization – at least in two dimensions: time and space.

As a conclusion, coordinators of interdisciplinary projects should have the opportunity to learn more about team building, conflict management, workshop techniques, project planning and administration.

3.1 Communication within an Interdisciplinary Project – Project Example

Janusz Holyst, Warsaw University of Technology (Poland), presented the CREEN project.

The objectives of CREEN are to develop new methods to recognize emerging critical events in evolving complex networks, coupled networks and active agent networks to apply these methods to the analysis of the emergence of new research topics scientific avalanches and the sudden emergence of crises in a social institution – the public trust in science.

The project focuses on social networks and particularly on the spreading of information in scientific and public communication networks.

CREEN tries to find out about the networks behind events and recommends presenting the results of scientific work in a way that everybody can understand them.

The project partners communicate via regular meetings, teleconferences, mutual meetings and publish all news and information on the projects website.

Basic project facts concerning communication within the project: Since the start of the project nine publications were established, two meetings were held, three teleconferences took place and four mutual visits. The project's website was launched under: www.creen.org.

3.2. Communication on an Interdisciplinary Project Outside the Scientific World – Project Example

Luc Leyns, from the Free University of Brussels (Belgium), presented the CONTROL CANCER STEM project.

The project's aims and approaches are to devise a novel therapy approach to eliminate cancer stem cells from solid tumours, by modulating their behaviour with appropriate signalling cues and therefore forcing them to differentiate into non-proliferative and non-malignant cells.

The project has experienced barriers with a view to internal communication.

These barriers evolved from the different scientific worlds (Biology, Biomathematics etc.) the project partners came from and the use of different tools and different (special-) languages.

Continuous communication between all partners was necessary to clarify several situations and furthermore the project partners identified the need to develop a common language. To reach this goal frequent contacts between the project partners is absolutely necessary (meetings, emails, phone). Furthermore it is important for each of the partners to get an insight to the discipline of the other partners.

Consortia have to keep in mind that if they intend to submit a proposal to the European Commission, the project partners need to establish a common language even before the start of the project in order to assure to have a common ground that can be built on.

Partners of interdisciplinary projects require open minds for different scientific backgrounds. They also have to raise awareness to the importance of not keeping the project outcomes to a specified group but to communicate results to the general public. The consortia furthermore need to be aware that they might need to solve issues concerning unpublished results, patenting and IPR.

According to Luc Leyns, implementation of interdisciplinarity occurs not only during the course of the project but from the first discussion, i.e. during the preparation of the project.

3.3. Presentation of www.interdisciplines.org

Gloria Origgi (CNRS), member of the NEST-IDEA expert group, presented an example of good practice of a communication tool for interdisciplinary research. The www.interdisciplines.org project is working with a software tool that allows the organization of interdisciplinary online conferences in cognitive and social sciences.

The final discussion clearly showed that there is a need within the scientific landscape to discuss issues of interdisciplinarity and the management of interdisciplinary research. The participants mainly raised issues concerning problems in establishing interdisciplinary consortia and how more awareness of the communication problems in such projects can be raised.

The presentations of the workshop as well as additional information on the NEST-IDEA-project can be viewed on www.nest-idea.net.

CHAPTER 43

WHEN DIVERSITY MEANS RICHNESS

Conclusion

MICHEL CLAESSENS

Information and Communication Unit, Directorate-General for Research, European Commission, 200 rue de la Loi, 1049 Brussels, Belgium, Tel +32-2-2959971, E-mail michel.claessens@ec.europa.eut

Abstract: Communicating research and engaging with the public about science is a key issue in Europe. Promoting and improving these activities is a key objective for scientists, research managers and educators but also for organisations involved in policy making, such as the European Commission. We are responsible for getting the record straight showing what science and research in Europe can really bring to European citizens

For sure, communicating research is not always easy. Nor does it come by magic. It takes education, patience, hard work – plus a high degree of professionalism. But the good news is that research organisations in Europe are now taking communication seriously. They are hiring professional science communicators to reach out to the public. At the same time, many European Union (EU) scientists are actively developing strong communication skills and activities.

As a result, science communication is now recognised as a genuine and valuable profession in Europe, as it has long been in the US. Far from being a secondary activity, sometimes detrimental to one's career, communication activities are increasingly integrated in the core business of being a scientist, although much progress is still required. In many countries, research and teaching are still the only factors that play a part in the career advancement of individual scientists.

1. UNDERSTANDING THE PUBLIC

Now, if scientists are encouraged or even obliged to *inform* about what they are doing, they also have an imperative to *listen*. Scientists these days must understand the social context within which they operate. What people worry about. What they

M. Claessens (ed.), Communicating European Research 2005, 243–246.

expect or need from science. What they do not want in their life. In short, the ivory tower is no longer an option.

Indeed, communicating is an imperative in a democracy, if one is to build trust and legitimacy in activities funded in great part by the public. It is also a simple question of common sense: there are so many exciting developments and the public should be informed about them.

A key question here is the following: Are Europeans really interested in science? The situation is a bit of a paradox. On the one hand 80% of them agree that science and technology "will improve the quality of life of future generations". Yet, at the same time, so few Europeans demonstrate a real interest in finding out more about science by reading articles and watching TV shows. Europeans seem to favour "passive information". More than merely putting science in the headlines we have a duty to reconnect European citizens to Europe, through science. This is a democratic imperative. This is also an economic imperative. In the global market for talent, we must attract the best talents for research and innovation, to keep Europe competitive in the knowledge economy.

What can we do to bring science and society closer together? There are two objectives at hand.

First, we should enhance the public recognition of the role of researchers in society. Second, we have to encourage researchers to acquire the necessary communication skills so that they in turn can inform society about their knowledge and discoveries.

2. EUROPEAN INITIATIVES AND EXPERTISE

The European Commission has recently launched a number of initiatives to foster a more professional approach to communicating science and technology.

Initiatives such as 2005's "Researchers in Europe" campaign saw thousands of people from universities, research centres and industry coming together with the general public all over Europe to communicate the real excitement a career in research can deliver.

European Commission's Science and Society programme has been particularly successful in supporting events that popularise science and bring science – and scientists – closer to the citizens. This includes science festivals, interactive exhibitions, and all kinds of exciting education and outreach initiatives for young people and the general public. This book shows that many initiatives and many organisations play a key role in this process. These have evolved from 'unidimensional' places to do 'hands on' experiments, to a far more ambitious role – that of multifaceted laboratories for all types of exciting innovations in science communication. For dialogue and discovery. For teaching and training. For wondering and dreaming.

Giving due recognition to effective science communication is also key. The EU Descartes Prizes for Science Communication held each year provides a perfect opportunity to showcase the achievements of EU researchers. It valorises all those – scientists, authors, journalists, TV producers – whose gift is to get people interested in science.

In addition to these initiatives, Europe promotes the sharing of expertises in this topic, through meetings, as happened at CER 2005, and through publications, websites etc. They make it possible to disseminate experiences and expertise throughout Europe, and the influx of other expertises from outside Europe. For example, at the CER 2005, scientists learned from training workshops pioneered in other parts of the world, such as *Science Meets the Parliament in Australia* or the *Congressional Visits Day in the United States of America*. Websites provide important information for scientists wanting to be more involved in science communication, such as the practical guidelines compiled by the European Commission, media contacts, statistics and documents that help when organising and evaluating events.

Through its Science and Society Action Plan, the European Commission promotes more and better communication of science and claims in particular that researchers, research organisations and industry have a specific responsibility vis-á-vis society in terms of providing scientific and technological information to Europe's citizens. We encourage those responsible for EU-funded projects to pay close attention to the 'public communication' dimension of their work. This is now a contractual obligation. We also provide support and practical assistance to project coordinators on how best to publicise the results of their work, illustrate its benefits to EU citizens, and, more generally demonstrate the unique added value of cooperation on a Europe-wide scale. Promoting a culture of science communication in the scientific community was also the objective of the CER 2005 conference.

The training of scientists in communication skills is an important part of that equation. European diversity is therefore a strength as it allows us to take advantage of different expertises and experiences existent in this matter in different countries in Europe. There are several initiatives in Europe, such as ESConet, a European funded network of people involved in science communication, which run workshops to equip scientists to communicate effectively with different audiences in a variety of scenarios, taking into account the cultural differences throughout Europe. They offer training in techniques for listening, dialogue and crisis management, as well as for communicating with different publics and provide new tools for the evaluation of its workshops. All these initiatives will increase and improve the contribution of scientists to the public discussion of Science & Technology (S&T) and their participation in communication activities.

There are at least two complementary ways for developing the participation of scientists in these activities:

- promoting the career of science communication professionals and
- rewarding scientists who participate in S&T communication activities.

These key ways still need to be further developed and supported in Europe.

3. NEW ACTIVITIES

All these activities will gain in momentum in the forthcoming Seventh Framework Programme 2007–2013. While the Programme will be adopted at the end of 2006, the European Commission has proposed to support several activities in relation to science communication, including to:

- bridge the gap between those who have a scientific education and those who do not;
- promote a taste for scientific culture in the direct neighbourhood of all citizens (calling upon cities, regions, foundations, science centres, museums, civil society organisations, etc.);

- provide an image of science and researchers which is meaningful to all, especially to young people;
- provide reliable and timely scientific information to the press and other media;
- support training actions to bridge the gap between the media and the scientific community;
- encourage a European dimension at science events targeting the public;
- promote science by audio-visual means via European co-productions and the circulation of science programmes;
- encourage a societal dialogue on research policy, and stimulate civil society organisations to become more involved in research activities;
- renew science communication, favouring modern means to achieve higher impact, helping scientists to work closely with media professionals;
- promote excellent trans-national research and science communication by the means of popular prizes;
- support research aimed at enhancing science communication in its methods and its products;
- support formal and informal science education in schools as well as through science centres and museums and other relevant means.

All these activities will improve the relationships between science and society and catalyse the reintegration of science in the European culture.